应用型本科院校计算机类专业校企合作实

基于Android的
手机应用软件开发教程

主　编　包依勤
副主编　陈　欢

南京大学出版社

应用型本科院校计算机类专业校企合作实训系列教材编委会

主 任 委 员：刘维周

副主任委员：张相学　徐　琪　杨种学(常务)

委　　　员(以姓氏笔画为序)：

　　　　王小正　王江平　王　燕　田丰春　曲　波

　　　　李　朔　李　滢　闵宇峰　杨　宁　杨立林

　　　　杨蔚鸣　郑　豪　徐家喜　谢　静　潘　雷

序 言

在当前的信息时代和知识经济时代,计算机科学与信息技术的应用已经渗透到国民生活的方方面面,成为推动社会进步和经济发展的重要引擎。

随着产业进步、学科发展和社会分工的进一步精细化,计算机学科新知识、新领域层出不穷,多学科交叉与融合的计算机学科新形态正逐渐形成。2012年,国家教育部公布的《普通高等学校本科专业目录(2012年)》中将计算机类专业分为计算机科学与技术、软件工程、网络工程、物联网工程、信息安全、数字媒体技术等专业。

随着国家信息化步伐的加快和我国高等教育逐步走向大众化,计算机类专业人才培养不仅在数量的增加上也在质量的提高上对目前的计算机类专业教育提出更为迫切的要求。社会需要计算机类专业的教学内容的更新周期越来越短,相应的,我国计算机类专业教育也将改革的目标与重点聚焦于如何培养能够适应社会经济发展需要的高素质工程应用型人才。

作为应用型地方本科高校,南京晓庄学院计算机类专业在多年实践中,逐步形成了陶行知"教学做合一"思想与国际工程教育理念相融合的独具晓庄特色的工程教育新理念。学生在社会生产实践的"做"中产生专业学习需求和形成专业认同,在"做"中增强实践能力和创新能力,在"做"中生成和创造新知识,在"做"中涵养基本人格和公民意识;同时学生应遵循工程教育理念,标准地"做",系统地"做",科学地"做",创造地"做"。

实训实践环节是应用型本科院校人才培养的重要手段之一,是应用型人才培养目标得以实现的重要保证。当前市场上一些实训实践教材导向性不明显,可操作性不强,系统性不够,与社会生产实际联系不紧密。总体上来说没有形成系列,同一专业的不同实训实践教材重复较多,且教材之间的衔接不够。

《教育部关于"十二五"普通高等教育本科教材建设的若干意见(教高[2011]05号)》要求重视和发挥行业协会和知名企业在教材建设中的作用,鼓励行业协会和企业利用其具有的行业资源和人才优势,开发贴近经济社会实际的教材和高质量的实践教材。南京晓庄学院计算机类专业积极开展校企联合实训实践教材建设工作,与国内多家知名企业共同规划建设"应用型本科院校计算机类专业校企合作实训系列教材"。

本系列教材是在计算机学科和计算机类专业课程体系建设基本成熟的基础上,参考《中国计算机科学与技术学科教程2002》(China Computing Curricula 2002,简称CCC2002)并借鉴ACM和IEEE CC2005课程体系,经过认真的市场调研,由我校优秀教学科研骨干和行业企业专家通力合作而成的,力求充分体现科学性、先进性、工程性。

本系列教材在规划编写过程中体现了如下一些基本组织原则和特点。

1. 贯彻了"大课程观"、"大教学观"和"大工程观"的教学理念。教材内容的组织和案例的甄选充分考虑了复杂工程背景和宏大工程视野下的工程项目组织、实施和管理,注重强化具有团队协作意识、创新精神等优秀人格素养的卓越工程师的培养。

2. 体现了计算机学科发展趋势和技术进步。教材内容适应社会对现代计算机工程人才培养的需求,反映了基本理论和原理的综合应用,反映了教学体系的调整和教学内容的及时更新,注重将有关技术进步的新成果、新应用纳入教材内容,妥善处理了传统知识的继承与现代工程方法的引进。

3. 反映了计算机类专业改革和人才培养需要。教材规划以2012年教育部公布的新专业目录为依据,正确把握了计算机类专业教学内容和课程体系的改革方向。在教材内容和编写体系方面注重了学思结合、知行合一和因材施教,强化了以适应社会需要为目标的教学内容改革,由知识本位转向能力本位,体现了知识、能力、素质协调发展的要求。

4. 整合了行业企业的优质技术资源和项目资源。教材采用校企联合开发和建设的模式,充分利用行业专家、企业工程师和项目经理的项目组织、管理、实施经验的优势,将企业实际实施的工程项目分解为若干可独立执行的案例,注重了问题探究、案例讨论、项目参与式教育教学方式方法的运用。

5. 突出了应用型本科院校基本特点。教材内容以适应社会需要为目标,突出"应用型"的基本特色,围绕培养目标,以工程应用为背景,通过理论与实践相结合,重视学生的工程应用能力的培养,增强学生的技能的应用。

相信通过这套"应用型本科院校计算机类专业校企合作实训系列教材"的规划出版,能够在形式上和内容上显著提高我国应用型本科院校计算机类专业实践教材的整体水平,继而提高计算机类专业人才的培养质量,培养出符合经济社会发展需要和产业需求的高素质工程应用型人才。

李洪天

南京晓庄学院　党委书记　教授

前　言

Android 一词的本义指"机器人"，同时也是 Google 于 2007 年 11 月 5 日正式发布的基于 Linux 平台的开源手机操作系统。该平台由操作系统、中间件、用户界面和应用软件组成，是首个为移动终端打造的真正开放和完整的移动软件。

《基于 Android 的手机应用软件开发教程》是计算机科学专业智能终端方向、嵌入式方向和网络工程方向的专业选修课，以培养移动互联网应用型人才为主。目前从 Android 人才需求类型来看，一类是偏向硬件驱动的，一类是偏向软件应用的。后者的需求更大，软件应用包括手机游戏、手机终端应用软件和其他手机应用软件的开发。本教材侧重理论和实践相结合，通过案例讲解手机应用软件开发的知识点，所有程序都可在 Android 手机上运行。

由于目前 Android 技术发展比较快，无论是相关书籍、培训还是大学教育，都处于初级阶段，因此 Android 人才短期内将供不应求。从长期来看，随着各种移动应用和手机游戏等内容需求日益增加，也将激励大中小型手机应用开发商加大对 Android 应用的开发力度，因此 Android 人才需求的潜力非常巨大。

本书在编写过程中，作者力求保持教学内容的系统性，由浅入深，深入浅出，同时以 Android 技术应用为主线，加入了组件编程技术、游戏编程技术、SQLite 数据库技术、图形编程技术、Google 地图、NDK 技术等内容，力求反映 Android 技术的最新发展成果。

在本书编写过程中，作者参考了南京多禾信息技术有限公司的项目案例，其中包括他们近年来积累的技术资料和程序代码，力求做到层次清楚，内容丰富，代码准确，具有实战性。这样，既便于读者循序渐进地系统学习，又能使读者深入了解到 Android 手机最新的编程技术，希望本书对读者掌握 Android 系统的开发技术有一定的帮助。

本书的第 1、2、3、4、5、6、7、8、9、10、11、12 章由包依勤执笔完成，第 13、14、15、16、17、18 章由陈欢执笔完成。全书由包依勤统稿。

在本书的编写过程中，扬种学老师、王小正老师对部分内容提出了建设性意见，同时在文档编辑和整理过程当中，王乐、谢磊等学生参与并做了大量的工作，在此谨表衷心的感谢。

限于作者的学术水平，错误与不妥之处在所难免，竭诚欢迎来自教师和学生有利于该教材改进的批评、意见和对文字或程序错误的指正。联系方式为：baoyiqin888@yahoo.com.cn

编　者
2012 年 12 月

目 录

前言 ·············· 1

第 1 章 Android 简介 ·············· 1
1.1 手机操作系统 ·············· 2
1.2 Android 起源 ·············· 2
1.3 Android 发展史 ·············· 4
1.4 Android 特征 ·············· 5
1.5 Android 体系结构 ·············· 5
习题与思考题 ·············· 6

第 2 章 Android 开发环境 ·············· 7
2.1 安装 Android 开发环境 ·············· 8
2.2 Android SDK ·············· 14
习题与思考题 ·············· 20

第 3 章 Android 应用程序 ·············· 21
3.1 第一个 Android 程序 ·············· 22
3.2 Android 程序结构 ·············· 26
3.3 命令行创建程序 ·············· 30
习题与思考题 ·············· 38

第 4 章 Android 生命周期 ·············· 39
4.1 程序生命周期 ·············· 40
4.2 Android 组件 ·············· 41
4.3 Activity 生命周期 ·············· 41
4.4 程序调试 ·············· 47
习题与思考题 ·············· 52

第 5 章 Android 用户界面 ·············· 53
5.1 用户界面基础 ·············· 54
5.2 界面控件 ·············· 55
5.3 界面布局 ·············· 66
5.4 菜单 ·············· 77
5.5 界面事件 ·············· 85
习题与思考题 ·············· 91

第 6 章 组件通信与广播消息 ·············· 93
6.1 Intent ·············· 94
6.2 广播消息 ·············· 106
习题与思考题 ·············· 109

第 7 章 Android 后台服务 ·············· 111
7.1 Service 简介 ·············· 112
7.2 本地服务 ·············· 113
7.3 远程服务 ·············· 129
习题与思考题 ·············· 146

第 8 章 对话框与提示信息 ·············· 147
8.1 对话框 ·············· 148
8.2 提示信息 Toast ·············· 151
8.3 温馨信息 Notification ·············· 153
习题与思考题 ·············· 155

第 9 章 Android 桌面组件 ·············· 157
9.1 AppWidget 框架类 ·············· 158
9.2 如何使用 Widget ·············· 158
9.3 Demo 讲解 ·············· 159

习题与思考题 …………………… 163

第 10 章　SQL 基础 ……………… 165
10.1　SQL 概述 …………………… 166
10.2　数据定义功能 ……………… 166
10.3　数据操纵功能 ……………… 170
10.4　查询功能 …………………… 171
习题与思考题 …………………… 180

第 11 章　数据存储和访问 ……… 181
11.1　简单存储 …………………… 182
11.2　文件存储 …………………… 187
11.3　数据库存储 ………………… 197
11.4　数据分享 …………………… 207
习题与思考题 …………………… 224

第 12 章　联系人 ………………… 227
12.1　联系人数据库 ……………… 228
12.2　对联系人的基本操作 ……… 229
习题与思考题 …………………… 232

第 13 章　Android 图形开发 …… 233
13.1　Drawable 对象 ……………… 234
13.2　Bitmap 对象 ………………… 237
13.3　Animation 对象 …………… 241
习题与思考题 …………………… 245

第 14 章　SurfaceView …………… 247
14.1　SurfaceView 简介 …………… 248
14.2　自定义 SurfaceView ………… 252

14.3　SurfaceView 的多线程 …… 254
习题与思考题 …………………… 259

第 15 章　2D 游戏开发 …………… 261
15.1　2D 游戏开发基础 …………… 262
15.2　简单游戏框架 ……………… 264
15.3　声音播放 …………………… 268
15.4　手势识别 …………………… 274
15.5　加速度传感器 ……………… 283
习题与思考题 …………………… 288

第 16 章　2D 游戏开发进阶 …… 289
16.1　游戏地图 …………………… 290
16.2　碰撞 ………………………… 290
16.3　游戏的状态控制 …………… 292
16.4　打砖块游戏实例 …………… 292
习题与思考题 …………………… 312

第 17 章　位置服务与地图 ……… 313
17.1　位置服务 …………………… 314
17.2　Google 地图应用 …………… 319
习题与思考题 …………………… 327

第 18 章　Android NDK 开发 …… 329
18.1　NDK 简介 …………………… 330
18.2　NDK 编译环境 ……………… 330
18.3　NDK 开发示例 ……………… 332
习题与思考题 …………………… 338

参考文献 …………………………… 339

第 1 章

Android 简介

☆1.1　手机操作系统
☆1.2　Android 起源
☆1.3　Android 发展史
☆1.4　Android 特征
☆1.5　Android 体系结构

本章学习目标：了解各种手机操作系统的特点；了解开放手机联盟的目的、组成和性质；了解 Android 平台的发展历史、特征和体系结构。

1.1 手机操作系统

目前手机操作系统主要有 6 种：

(1) Windows Mobile：由微软公司推出的移动设备操作系统，支持播放音视频、浏览网页、MSN 聊天和收发电子邮件等功能，缺点是硬件配置高、耗电量大、电池续航时间短、硬件成本高。

(2) PalmOS：由 3Com 公司的 Palm Computing 开发，32 位嵌入式操作系统，所占的内存小，现已被 HP 公司收购。

(3) Symbian：由 Symbian 开发和维护，后被诺基亚收购。Symbian 是一个实时多任务的 32 位操作系统，功耗低、内存占用少、应用界面框架灵活，不开放核心代码，但公开 API 文档。

(4) Linux：由计算机 Linux 操作系统演变而来，开放源代码，有一些先天不足。

(5) iPhoneOS：由苹果公司开发的操作系统。iPhoneOS 的系统构架，支持内置加速器。

(6) Android：由谷歌发布的基于 Linux 的开源手机平台。Android 系统是第一个完全定制、免费、开放的手机平台，使用 Java 语言开发，支持多种硬件设备。

1.2 Android 起源

开放手机联盟(Open Handset Alliance，OHA)是由谷歌公司于 2007 年发起的一个全球性的联盟组织，成立时包含 34 家联盟成员，现在已经增加到 50 家。开放手机联盟的组成如图 1-1 所示。

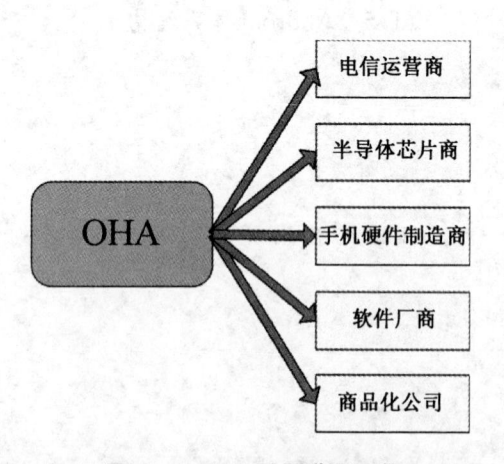

图 1-1 OHA 手机联盟组成

联盟组织的目标：研发移动设备的新技术，用以大幅削减移动设备开发与推广的成本，同时通过联盟各个合作方的努力，建立移动通信领域新的合作环境，促进创新移动设备的开发，开创目前移动平台实现的用户体验。

电信运营商：中国移动通信、KDDI（日本）、NTT DoCoMo（日本）、Sprint Nextel（美国）、T‑Mobile（美国）、Telecom（意大利）、中国联通、SoftBank（日本）、Telefonica（西班牙）和 Vodafone（英国）。如图1-2所示。

图1-2　电信运营商

半导体芯片商：Audience（美国）、AKM（日本）、ARM（英国）、Atheros Communications（美国）、Broadcom（美国）、Intel（美国）、Marvell（美国）、nVIDIA（美国）、Qualcomm（美国）、SiRF（美国）、Synaptics（美国）、ST‑Ericsson（意大利、法国和瑞典）和 Texas Instruments（美国）。如图1-3所示。

图1-3　半导体芯片商

手机硬件制造商：鸿碁（中国台湾）、华硕（中国台湾）、佳明（中国台湾）、宏达电（中国台湾）、LG（韩国）、三星（韩国）、华为（中国）、摩托罗拉（美国）、索尼爱立信（日本和瑞典）和东芝（日本）。如图1-4所示。

图1-4　手机硬件制造商

软件厂商：Ascender Corp（美国）、eBay（美国）、Coogle（美国）、Living Image（日本）、Nuance Communications（美国）、Myraid（瑞士）、Omron（日本）、PacketVideo（美国）、SkyPop（美国）、Svox（瑞士）和 SONiVOX（美国）。如图1-5所示。

图1-5　软件厂商

商品化公司：Aplix Corporation（日本）、Noser Engineering（瑞士）、Borqs（中国）、TAT—The Astonishing（瑞典）、Teleca AB（瑞典）和 Wind River（美国）。如图1-6所示。

图1-6　商品化公司

1.3　Android 发展史

2007年11月5日，开放手机联盟成立。2007年11月12日，谷歌发布 Android SDK 预览版，这是第一个对外公布的 Android SDK，为发布正式版收集用户反馈。如图1-7所示。

图1-7　收集用户反馈　　　　图1-8　第一款 Android 手机

2008年4月17日，谷歌举办开发者竞赛。

2008年8月28日，谷歌开通 Android Market，供 Android 手机下载需要使用的应用程序。

2008年9月23日，发布 Android SDK v1.0版，这是第一个稳定的 SDK 版本。

2008年10月21日，谷歌开放 Android 平台的源代码。

2008年10月22日，第一款 Android 手机 T-Mobile G1 在美国上市，由中国台湾的宏达电制造。如图1-8所示。

2009年2月，发布 Android SDK v1.1版。

2009年2月17日，第二款 Android 手机 T-Mobile G2 正式发售，仍由中国台湾的宏达电制造。如图1-9所示。

2009年4月15日，发布 Android SDK v1.5版。

2009年6月24日，中国台湾的宏达电发布了第三款 Android 手机 HTC Hero。如图1-10所示。

图1-9　第二款 Android 手机

图 1-10　第三款 Android 手机

1.4　Android 特征

Android 提供访问硬件的 API 函数,简化摄像头、GPS 等硬件的访问过程,具有自己的运行时环境和虚拟机;提供丰富的界面控件供使用者之间调用,加快用户界面的开发速度,保证 Android 平台上程序界面的一致性;提供轻量级的进程间通讯机制 Intent,使跨进程组件通信和发送系统级广播成为可能;提供了 Service 作为无用户界面、长时间后台运行的组件,支持高效、快速的数据存储方式。

1.5　Android 体系结构

采用软件堆层的架构,共分为四层,如图 1-11 所示。

图 1-11　Android 体系结构

Linux 内核:硬件和其他软件堆层之间的一个抽象隔离层,提供安全机制、内存管理、进程管理、网络协议堆栈和驱动程序等。

中间件层:由函数库和 Android 运行时构成。函数库主要提供一组基于 C/C++的函数库;Surface Manager,支持显示子系统的访问,提供应用程序与 2D、3D 图像层的平滑连接;Media Framework,实现音视频的播放和录制功能;SQLite,轻量级的关系数据库引擎;OpenGL ES,基于 3D 的图像加速;FreeType,位图与矢量字体渲染;WebKit,Web 浏览器引擎;SGL,2D 图像引擎;SSL,数据加密与安全传输的函数库;Libc,标准 C 运行库,Linux 系统中底层应用程序开发接口;Android 运行时环境中,核心库提供 Android 系统的特有函数功能和 Java 语言函数功能,Dalvik 虚拟机实现基于 Linux 内核的线程管理和底层内存管理。

应用程序框架:提供 Android 平台基本的管理功能和组件重用机制。Activity Manager,管理应用程序的生命周期;Windows Manager,启动应用程序的窗体;Content Provider,共享私有数据,实现跨进程的数据访问;Package Manager,管理安装在 Android 系统内的应用程序;Teleghony Manager,管理与拨打和接听电话的相关功能;Resource Manager,允许应用程序使用非代码资源;Location Manager,管理与地图相关的服务功能;Notification Manager,允许应用程序在状态栏中显示提示信息。

应用程序:提供一系列的核心应用程序,包括电子邮件客户端、浏览器、通讯录和日历等。

习题与思考题

1. 简述 6 种主流的手机操作系统的特点。
2. 简述 Android 平台的特征。
3. 描述 Android 平台体系结构的层次划分,并说明各个层次的作用。

第 2 章

Android 开发环境

☆2.1 安装 Android 开发环境
☆2.2 Android SDK

本章学习目标：掌握 Android 开发环境的安装配置方法；了解 Android SDK 的目录结构和示例程序；了解各种 Android 开发工具的用途。

2.1 安装 Android 开发环境

2.1.1 安装 JDK 和 Eclipse

1. 确认安装 JRE

JRE 即 Java 运行环境，在没有安装 JRE 时打开 Eclipse，会出现如图 2-1 所示的错误提示。JDK 中包含 JRE，所以只需下载 JDK。

图 2-1 Eclipse 错误提示

2. 安装 JDK

用浏览器打开网页 http://java.sun.com/javase/downloads/index.jsp，选择下载：JDK6 Update 14。如图 2-2 所示。

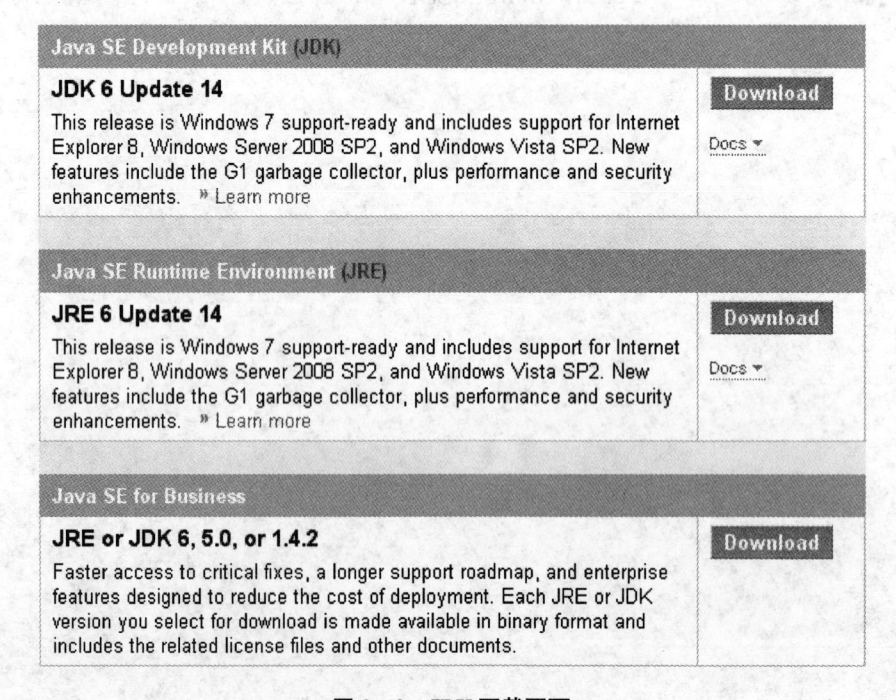

图 2-2 JDK 下载页面

运行平台选择:Windows,语言选择:Multi-language。如图 2-3 所示。

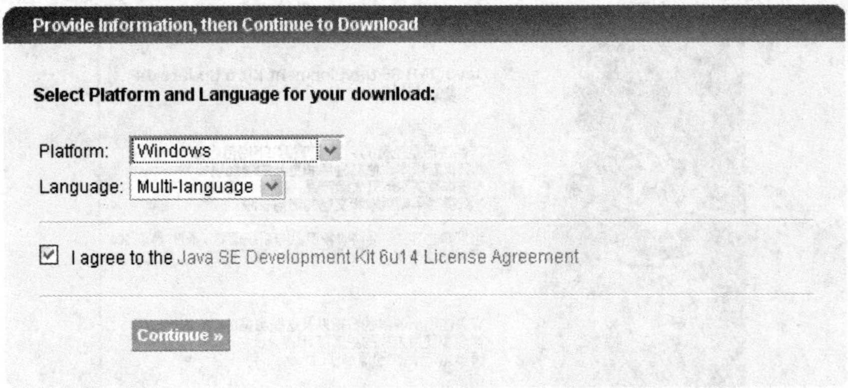

图 2-3　语言选择页面

下载 jdk-6u14-windows-i586.exe。如图 2-4 所示。

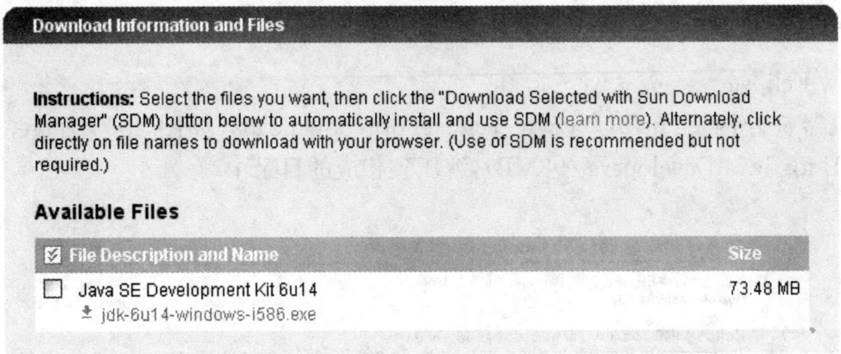

图 2-4　JDK 选择页面

选择安装目录:C:\Program Files\Java\jdk1.6.0_14\。如图 2-5 所示。

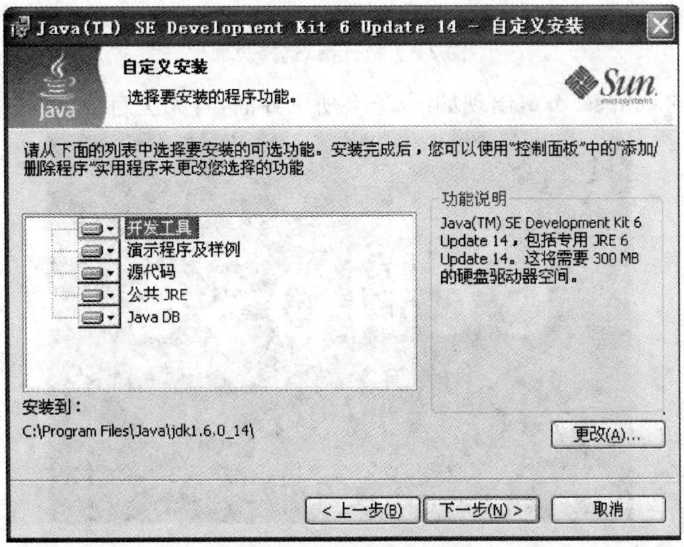

图 2-5　安装目录选择页面

安装成功，出现如图 2-6 所示界面。

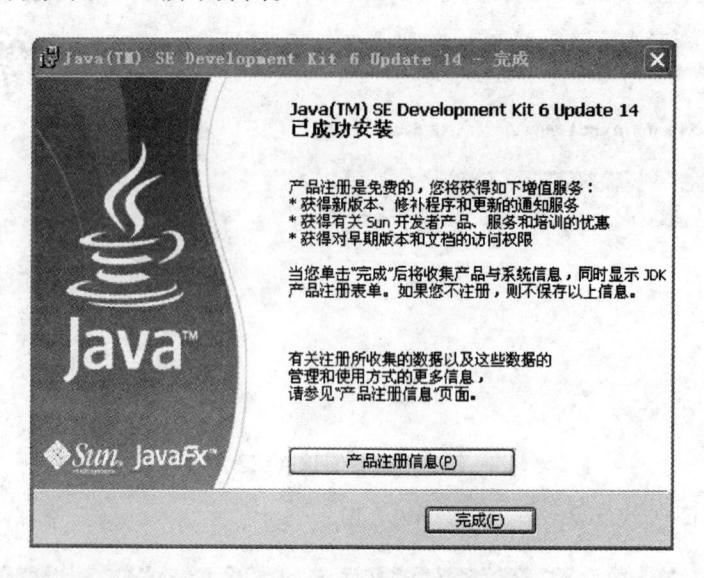

图 2-6 安装成功页面

3. 启动 Eclipse

用浏览器打开网页：http://www.eclipse.org/downloads，如图 2-7 所示，选择下载：Eclipse IDE for Java Developers (92MB)，解压到相应的目录中。

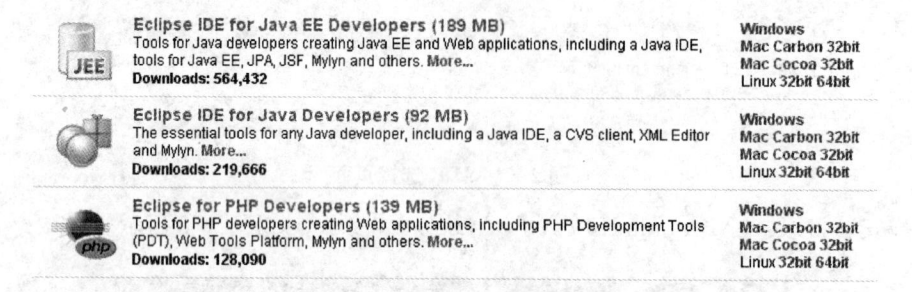

图 2-7 eclipse 选择页面

双击目录中的 eclipse.exe，出现如图 2-8 所示界面，若无法启动，尝试重启机器。

图 2-8 eclipse 启动页面

选择工作目录：E:\Android\workplace，建议选择复选框，将工作目录设成默认工作目录。如图 2-9 所示。

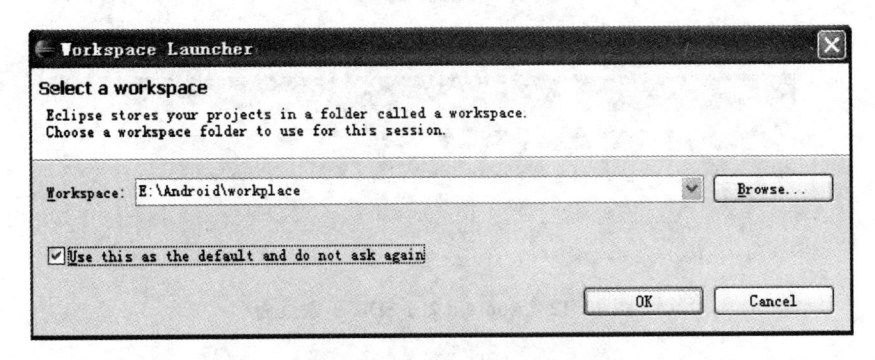

图 2-9　工作目录选择界面

确认后出现 Eclipse 集成开发环境如图 2-10 所示。

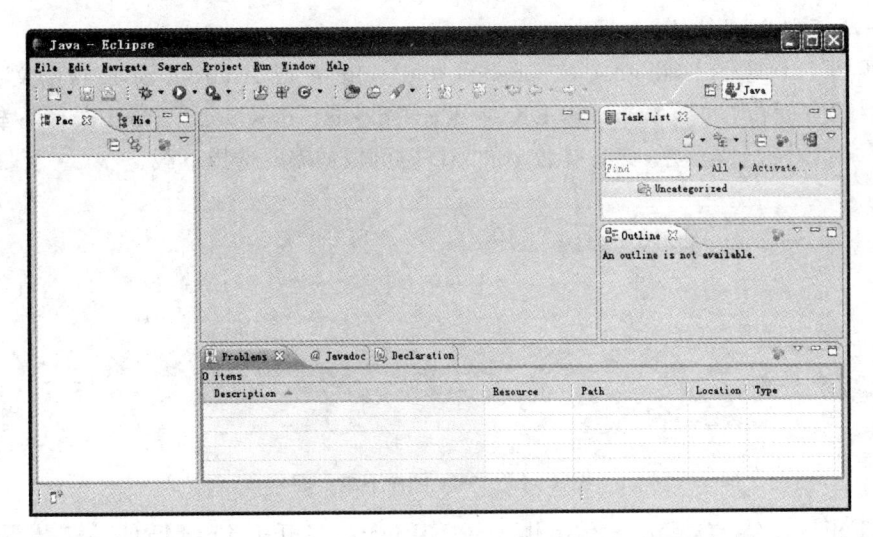

图 2-10　Eclipse 集成开发环境

2.1.2　安装 Android SDK

Android 开发站点：http://code.google.com/android，点击"下载 SDK"。如图 2-11 所示。

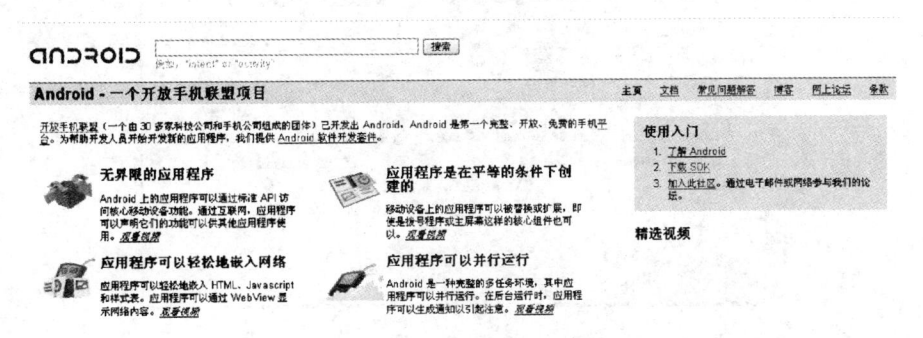

图 2-11　Android SDK 下载页面

如图 2-12 所示，选择 Windows 版本的 Android SDK，Android SDK 只需解压，无需安装。

图 2-12　Android 2.1 SDK 下载页面

2.1.3　安装 ADT 插件

安装 ADT 有两种方法。

1. 手动下载 ADT 插件

下载网址：http://dl-ssl.google.com/android/eclipse/或 Android SDK 帮助文档中的下载页面，下载页面位置：＜Android SDK＞/docs/sdk/adt_download.html，选择下载：ADT_0.9.1.zip。如图 2-13 所示，如有高版本的 ADT，可根据需要选择下载。

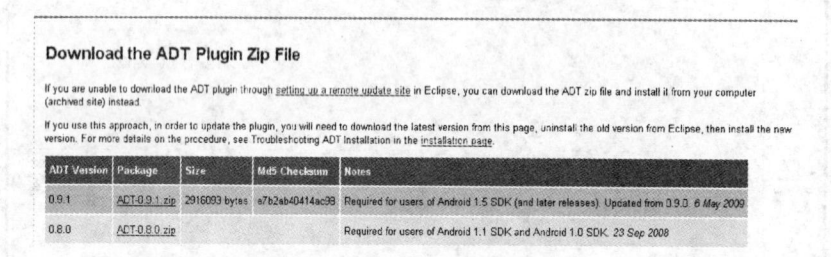

图 2-13　ADT 插件下载页面

启动 Eclipse，选择 Help → Install New Software，打开 Eclipse 插件的安装界面，点击 "Add"。如图 2-14 所示。

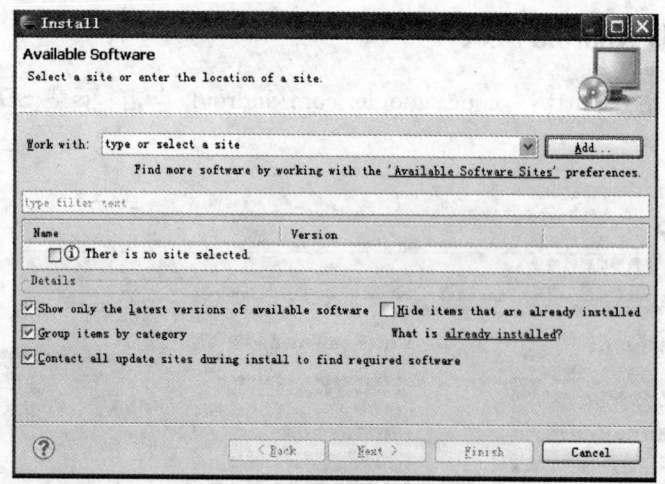

图 2-14　安装插件页面

点击 Archive,选择 ADT 保存位置。如图 2-15 所示。

图 2-15　选择 ADT 保存位置界面

选择 ADT 插件安装选项:复选 Android DDMS 和 Android Development Tools。如图 2-16 所示。

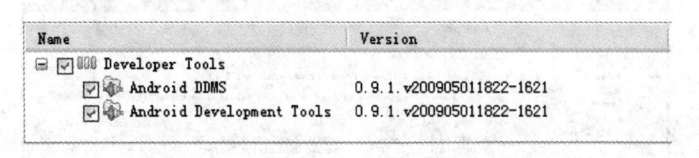

图 2-16　选择 ADT 插件安装界面

认可 ADT 插件的许可证(在 ADT 安装前),如图 2-17 所示。

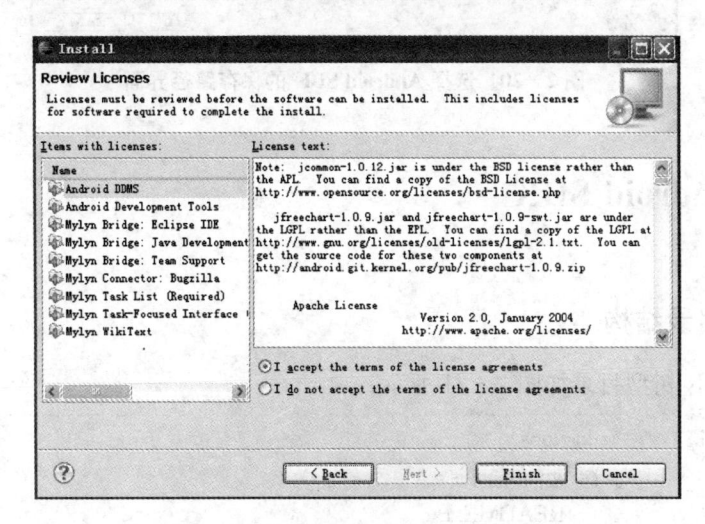

图 2-17　认可 ADT 插件的许可证界面

提示重启 Eclipse,即完成下载,如图 2-18 所示。

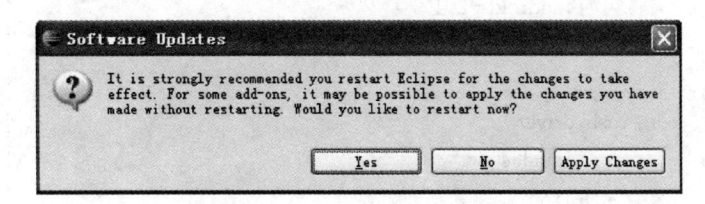

图 2-18　完成下载选择界面

2. 自动下载 ADT 插件

自动下载 ADT 插件的方法和手动方法安装相似,不同之处在于:直接输入插件压缩包的下载地址 http://dl—ssl.google.com/android/eclipse/。如图 2-19 所示，完成下载。

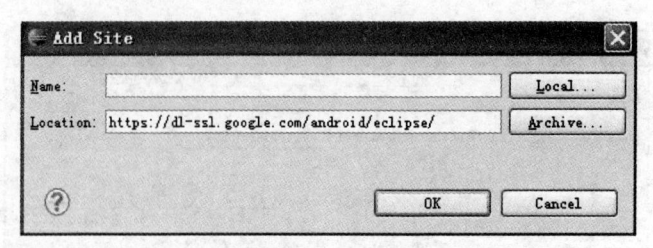

图 2-19　自动下载 ADT 插件选择界面

设置 Android SDK 的保存路径，选择 Windows→Preferences 打开 Eclipse 的配置界面，输入 Android SDK 的保存路径，点击"Apply"，如图 2-20 所示。

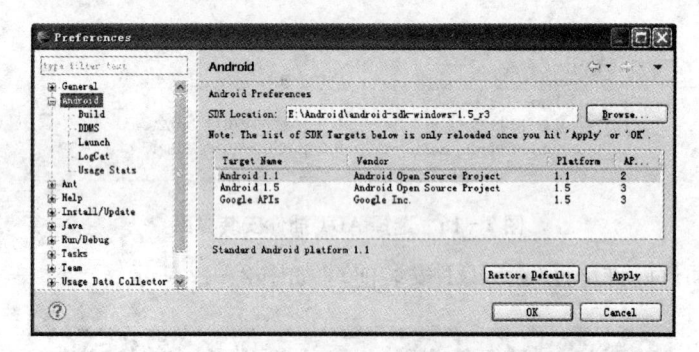

图 2-20　选择 Android SDK 的保存路径界面

2.2　Android SDK

2.2.1　目录结构

Android SDK 树型目录如图 2-21 所示。

```
        (+)add-ons
            (+)google_apis-3
            (-)README.txt
    (+)docs
    (+)platforms
        (+)android-2.1
        (+)android-2.2
    (+)tools
    (+)usb_driver
        (+)amd64
        (+)x86
    (-)documentation.html
    (-)RELEASE_NOTES.html
```

图 2-21　Android SDK 树型目录

add-ons 目录下的是 **Google** 提供地图开发的库函数,支持基于 **Google Map** 的地图开发。

docs 目录下的是 **Android SDK** 的帮助文档,通过根目录下的 **documentation.html** 文件启动。

platforms 目录中存在两个子目录 android-2.1 和 android-2.2,分别用来保存 **2.1** 版本和 **2.2** 版本的 **Android SDK** 的库函数、外观样式、程序示例和辅助工具等。

tools 目录下的是通用的 **Android** 开发和调试工具。

usb_driver 目录下保存了用于 **amd64** 和 **x86** 平台的 **USB** 驱动程序。**RELEASE_NOTES.html** 是 **Android SDK** 的发布说明。

2.2.2 示例程序

如图 2-22 所示为 HelloActivity 界面示例。

图 2-22　HelloActivity 界面　　　　图 2-23　ApiDemos 和 App/Alarm 界面

如图 2-23 所示为 ApiDemos 和 App/Alarm 示例界面。如图 2-24 所示为 SkeletonApp 示例界面,如图 2-25 所示为 NotesList 示例界面。

图 2-24　SkeletonApp 界面 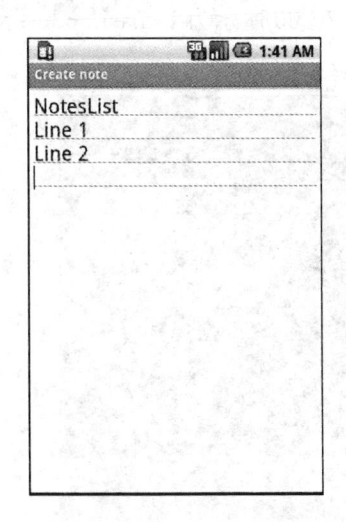　　　　图 2-25　NotesListp 界面

如图 2-26 所示为 Home 选择示例界面,如图 2-27 所示为 Home 示例界面。

图 2-26　Home 选择界面　　　　图 2-27　Home 界面

如图 2-28 所示为 SoftKeyboar 示例界面，如图 2-29 所示为 Snake 示例界面。

图 2-28　Softkeyboar 界面　　　　图 2-29　Snake 界面

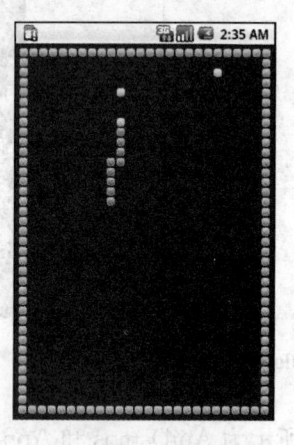

如图 2-30 所示为 LunarLander 示例界面，如图 2-31 所示为 JetBoy 示例界面。

图 2-30　LunarLander 选择界面　　　　图 2-31　JetBoy 界面

2.2.3 开发工具

Android 模拟器是 Android SDK 最重要的工具,支持加载 SD 卡映像文件,更改模拟网络状态、延迟和速度,模拟电话呼叫和接收短信等,不支持接听真实电话,USB 连接,摄像头捕获,设备耳机,电池电量和 AC 电源检测,SD 卡插拔检查和使用蓝牙设备。Android 模拟器外观如图 2-32 所示。

外　观	外观 ID	说　明
	HVGA-L	解析度:480×320 方向:横向
	HVGA-P	解析度:320×480 方向:纵向 缺省配置

图 2-32　Android 模拟器外观

1. Android 调试器

连接 Android 设备和模拟器的工具,客户端/服务器程序包含守护程序、服务器程序和客户端程序。

2. DDMS

Android 系统中内置的调试工具,监视 Android 系统中的进程和堆栈信息,查看 logcat 日志,实现端口转发服务和屏幕截图,支持模拟电话呼叫和 SMS 短信以及浏览 Android 模拟器文件系统等功能。

启动文件:<Android SDK>/tools/ddms.bat,调试 DDMS:Windows→Open Perspective→DDMS。

打开 Show View 的选择对话框:Windows→Show View→Other,打开 Show View 的选择对话框。如图 2-33 所示。

图 2-33　Show View 界面

DDMS 中的设备管理器可同时检控多个 Android 模拟器，显示每个模拟器中所有正在运行的进程，提供屏幕截图功能。如图 2-34 所示。

图 2-34 模拟器进程界面

DDMS 中的模拟器控制器可以控制 Android 模拟器的网络速度和延迟，模拟语音和 SMS 短信通信。如图 2-35 所示。

网络速率支持 GSM、HSCSD、GPRS、EDGE、UMTS、HSDPA 制式和全速率。

网络延迟支持 GPRS、EDGE、UMTS 制式和无延迟。

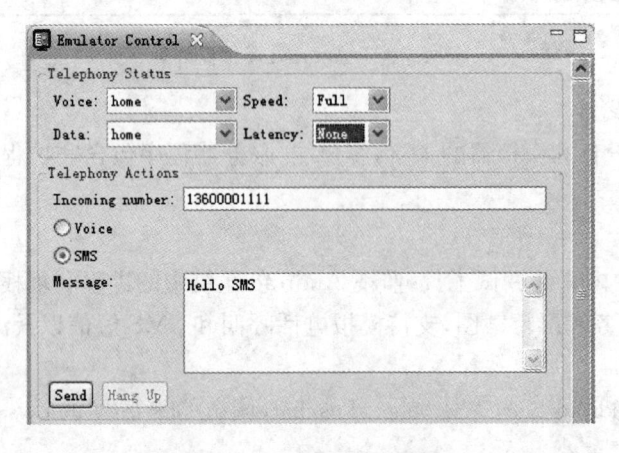

图 2-35 模拟器控制界面

如图 2-36 所示为电话呼入显示界面，如图 2-37 所示为 SMS 短信显示界面。

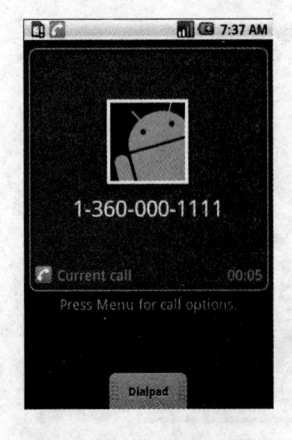

图 2-36 电话呼入界面　　　　图 2-37 SMS 短信界面

如图 2-38 所示为 DDMS 中的文件浏览器，支持上传、下载和删除 Android 内置存储器上的文件，显示文件和目录的名称、权限、建立时间等。

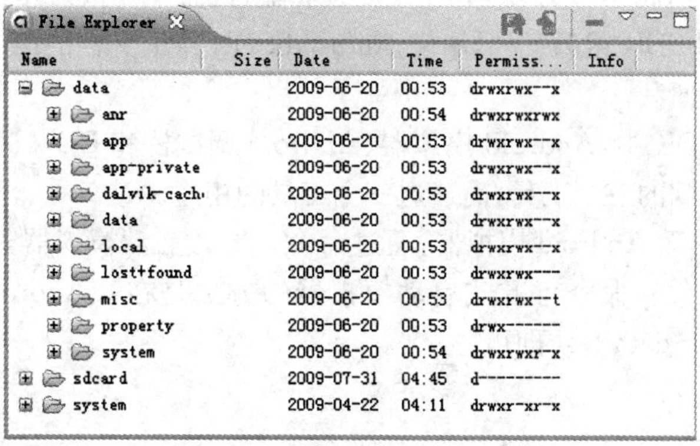

图 2-38　DDMS 中的文件浏览器

如图 2-39 所示为 DDMS 中的日志浏览器，用于浏览 Android 系统、Dalvik 虚拟机或应用程序产生的日志信息，有助于快速定位应用程序产生的错误。

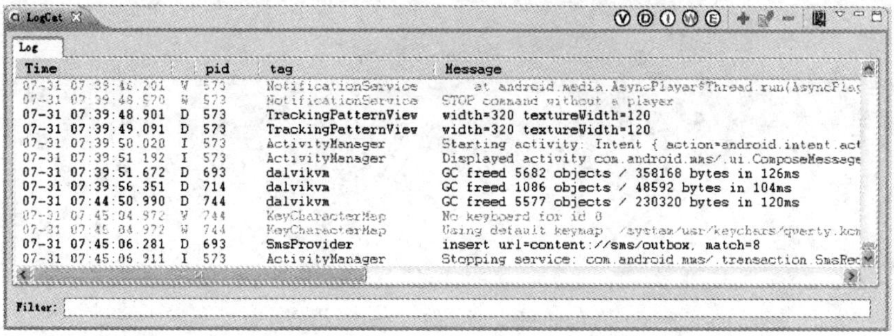

图 2-39　DDMS 中的日志信息

Android SDK 中的其他工具如表 2-1 所示。

表 2-1　android 开发工具

工具名称	启动文件	说　明
数据库工具	sqlite3.exe	用来创建和管理 SQLite 数据库
打包工具	apkbuilder.bat	将应用程序打包成 apk 文件
层级观察器	hierarchyviewer.bat	对用户界面进行分析和调试，以图形化的方式展示树形结构的界面布局
跟踪显示工具	traceview.bat	以图形化的方式显示应用程序的执行日志，用来调试应用程序，分析执行效率
SD 卡映像创建工具	mksdcard.exe	建立 SD 卡的映像文件
NinePatch 文件编辑工具	draw9patch.bat	NinePatch 是 Android 提供的可伸缩的图形文件格式，基于 PNG 文件。draw9patch 工具可以使用 WYSIWYG 编辑器建立 NinePatch 文件

习题与思考题

1. 尝试安装 Android 开发环境,并记录安装和配置过程中所遇到的问题。

2. 浏览 Android SDK 帮助文档,了解 Android SDK 帮助文档的结构和用途,这样会对以后的学习带来极大的便利。

3. 在 Android SDK 中,Android 模拟器、Android 调试桥和 DDMS 是 Android 应用程序开发过程中经常使用到的三个工具,简述这三个工具的用途。

4. 为了进一步熟悉 Android 模拟器,通过命令行方式启动模拟器,并在模拟器中尝试使用各种功能和应用软件。命令行方式启动模拟器的方法是在＜Android SDK＞/tools 目录中,输入命令 emulator-data test 即可。

第 3 章

Android 应用程序

☆3.1 第一个 Android 程序
☆3.2 Android 程序结构
☆3.3 命令行创建程序

本章学习目标：掌握使用 Eclipse 开发 Android 应用程序的方法；掌握 Android 虚拟设备的创建方法，了解 R.java 文件的用途和生成方法，了解 AndroidManifest.xml 文件的用途；了解 Android 的程序结构，了解使用命令行创建 Android 应用程序方法。

3.1 第一个 Android 程序

创建 Android 工程：在 Eclipse 程序中建立过的 Android 工程，工程名称和目录结构将显示在 Package Explorer 区域内，如图 3-1 所示。

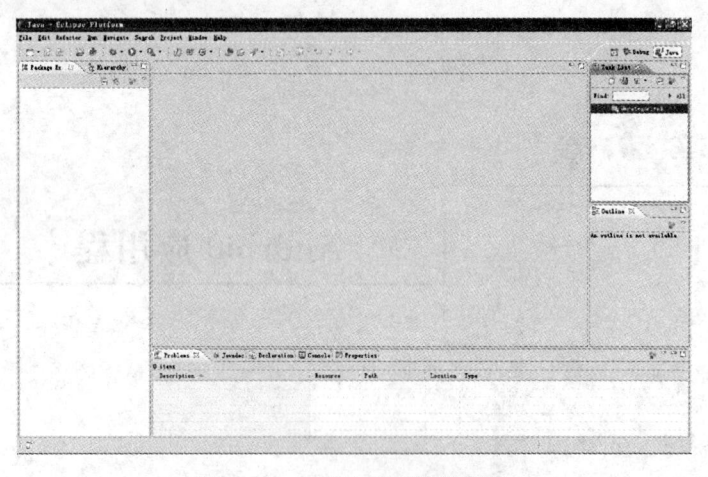

图 3-1　Eclipse 环境的展示

打开 Android 工程向导：File→New→Project…│Android→Android Project 或 File→New→Other…│Android→Android Project，如图 3-2 所示。

图 3-2　Android 工程向导

填写工程名称：工程名称必须唯一，不能与已有的工程重名，在 Project name 中填入 HelloAndroid，如图 3-3 所示。

图 3-3　填写工程名称

选择创建方式：可以创建新 Android 工程，也可利用已有代码创建 Android 工程，缺省为创建新 Android 工程；选择"Create new project in workspace"，可以使用默认位置存储，也可取消复选框，选择其他位置保存，缺省为使用默认位置 E:/Android/workplace/。使用默认位置存储将保存在：E:/Android/workplace/HelloAndroid，如图 3-4 所示。

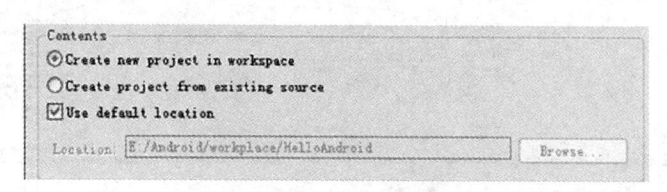

图 3-4　选择创建方式

选择编译目标：选择不同版本的 Android 系统，引入不同版本的 android.jar 包。这里选择标准的 1.5 版本 Android 系统，也可选择更高版本的 Android 系统，如图 3-5 所示。

图 3-5　选择编译目标示例

填写相关信息：应用程序名称，即 Android 程序在手机中显示的名称，显示在手机的顶部。在 Application name 填入 HelloAndroid，包名称是包的命名空间，需遵循 Java 包的命名方法，由两个或多个标识符组成，中间用点隔开。为了包名称的唯一性，可以采用反写电子邮件地址的方式。在 Package name 填入 edu.xzceu.HelloAndroid。

创建 Activity 是个可选项，如需要自动生成一个 Activity 的代码文件，则选择该项。Activity 的名称与应用程序的名称不同，但为了简洁，可以让他们相同，表示这个 Activity 是 Android 程序运行时首先显示给用户的界面。在 Create Activity 填入 HelloAndroid。

SDK 最低版本是 Android 程序能够运行的最低的 API 等级，如果手机的 API 等级低于程序的 SDK 最低版本，则程序无法在该 Android 系统中运行。在这里，之前选择标准的 1.5 版本的 Android 系统，则 SDK 等级被自动填入 3，此项无需更改，如图 3-6 所示。

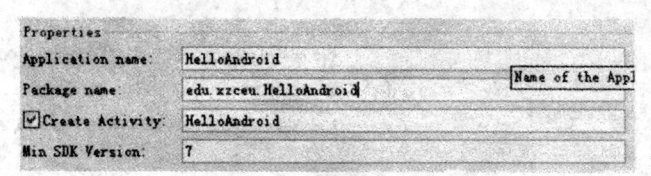

图 3-6　信息填写示例

启动 Eclipse，如图 3-7 所示。

图 3-7 android 工程创建完成

Android 虚拟设备（AVD）是对 Android 模拟器进行自定义的配置清单，能够配置 Android 模拟器的硬件列表和外观，支持 Android 系统版本、附件 SDK 库和储存设置等信息。因为 1.5 版本的 Android SDK 中没有附带任何配置好的 AVD，所以需建立一个 AVD。

建立 AVD，使用 Windows 系统命令行工具 CMD：开始→运行→CMD 中启动命令行工具，并进入＜Android SDK＞/tools 目录下。在这里，Android SDK 安装在 E:\Android\android-sdk-windows-1.5_r1，利用 android list targets 命令搜索＜Android SDK＞/platforms 和＜Android SDK＞/add-ons 目录下所有有效的 Android 系统映像，并将 Android 系统映像列表显示在命令行工具中，共有三个可以选择的编译目标：1.1 版本 Android 系统、1.5 版本 Android 系统和 Google API 的 1.5 版本 Android 系统，其中前两个系统映像的 Type 属性是 Platform，自行配置模拟器的硬件配置清单，最后一个的 Type 属性是 Add-On，不能更改配置清单，如图 3-8 所示。

图 3-8 Android 系统映像

使用 android create avd-n android1.5-t 2 命令，以 id 为 2 的 1.5 版本 Android 系统为目标，建立一个名为 Android1.5 的 AVD。-n 参数表明 AVD 的名称，-t 参数表明选择的 Android 系统映像的 id 值，可以直接缺省的硬件配置，当然也可以重新定制模拟器支持的硬件清单，如图 3-9 所示。

图 3-9　建立 AVD 模拟器示例

在建立过程中，Android 工具会在文件系统中建立 Android1.5.ini 文件和 Android1.5.avd 目录。Android1.5.ini 文件用来保存 Android1.5.avd 目录所在的位置，Android1.5.avd 目录用来保存 AVD 配置文件、用户数据文件、SD 卡映像和其他模拟器运行过程中可能产生的文件。如果用户使用的是 Windows XP 系统，则目录保存在 C:\Documents and Settings\<user>\.android\下，如果用户使用的是 Windows Vista/7 系统，则目录保存在 C:\Users\<user>\.android 下。

启动选项，路径：Run→Run Configuration 或 Run→Debug Configuration 配置模拟器的启动选项，可以选择不同的 AVD、配置网络速度、网络延迟、控制台的字符编码和标准输入输出等内容。一般只需选择正确的 AVD 即可。

启动 Android 程序，路径：Run→Run | Android Application 或 Run→Debug | Android Application，在程序调试完毕后，可直接再次运行 Android 程序，以便节约启动模拟器的时间，如图 3-10 所示。

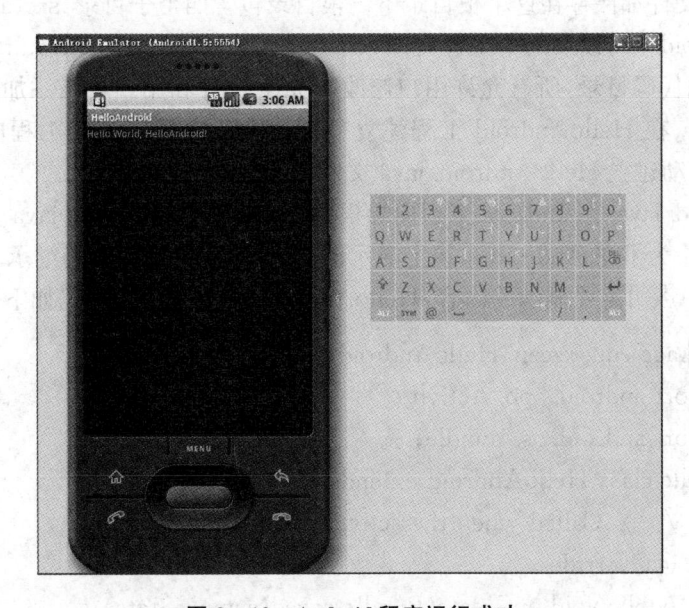

图 3-10　Android 程序运行成功

3.2 Android 程序结构

在建立 HelloAndroid 程序的过程中，ADT 会自动建立一些目录和文件，这些目录和文件有其固定的作用，有的允许修改，有的不能修改。Android 程序目录结构如图 3-11 所示，下面逐一的介绍。

图 3-11 Android 程序目录结构

在"Package Explore"中，ADT 以工程名称 HelloAndroid 作为根目录，将所有字段生成的和非自动生成的文件都保存在这个根目录下。根目录包含四个子目录：src、assets、res 和 gen，一个库文件 android.jar，以及两个工程文件 Androidmanifest.xml 和 default.properties。

src 目录是源代码目录，所有允许用户修改的 java 文件和用户自己添加的 java 文件都保存在这个目录中。在 HelloAndroid 工程建立初期，ADT 根据用户在工程向导中的"Create Activity"选项，自动建立 HelloAndroid.java 文件，如图 3-11 所示。

HelloAndroid.java 是 Android 工程向导根据 Activity 名称创建的 java 文件，这个文件完全可以手工修改。为了在 Android 系统上显示图形界面，需要使用代码继承 Activity 类，并在 onCreate() 函数中声明需要显示的内容，HelloAndroid.java 文件的代码如下。

```
1. package edu.xzceu.HelloAndroid；
2. import android.app.Activity；
3. import android.os.Bundle；
4. public class HelloAndroid extends Activity {
5.     /** Called when the activity is first created. */
6.     @Override
7.     public void onCreate(Bundle savedInstanceState) {
8.         super.onCreate(savedInstanceState)；
```

```
9.        setContentView(R.layout.main);  }
10. }
```

第 2 行和第 3 行的代码通过 android.jar 从 Android SDK 中引入了 Activity 和 Bundle 两个重要的包，用以子类继承和信息传递。第 4 行代码声明 HelloAndroid 类继承 Activity 类。第 8 行代码表明需要重写 onCreate() 函数。第 7 行代码的 onCreate() 会在 Activity 首次启动时被调用，为了便于理解，可以认为 onCreate() 是 HelloAndroid 程序的主入口函数，第 8 行代码调用父类的 onCreate() 函数，并将 savedInstanceState 传递给父类，savedInstanceState 是 Activity 的状态信息。第 9 行代码声明了需要显示的用户界面，此界面是用 XML 语言描述的界面布局，保存在 scr/layout/main.xml 中。

gen 目录是 1.5 版本新增的目录，用来保存 ADT 自动生成的 java 文件，例如 R.java 或 AIDL 文件。R.java 文件是 ADT 自动生成的文件，包含对 drawable、layout 和 values 目录内的资源的引用指针，Android 程序能够直接通过 R 类引用目录中的资源。

R.java 文件不能手工修改，如果向资源目录中增加或删除了资源文件，则需要在工程名称上右击，选择 Refresh 来更新 R.java 文件中的代码。R 类包含的几个内部类，分别与资源类型相对应，资源 ID 便保存在这些内部类中。例如子类 drawable 表示图像资源，内部的静态变量 icon 表示资源名称，其资源 ID 为 0x7f020000。一般情况下，资源名称与资源文件名相同。

HelloAndroid 工程生成的 R.java 文件的代码如下。

```
1. package edu.xzceu.HelloAndroid;
2. public final class R {
3.     public static final class attr {
4.     }
5.     public static final class drawable {
6.         public static final int icon=0x020000;
7.     }
8.     public static final class layout {
9.         public static final int main=0x030000;
10.    }
11.    public static final class string {
12.        public static final int app_name=0x040001;
13.        public static final int hello=0x040000;
14.    }
15. }
```

引用资源，资源引用有两种情况：一种是在代码中引用资源；另一种是在资源中引用资源代码来引用资源。需要使用资源 ID 时可以通过[R.resource_type.resource_name]或[android.R.resource_type.resource_name]获取，resource_type 代表资源类型，也就是 R 类中的内部类名称，resource_name 代表资源名称，对应资源的文件名或在 XML 文件中定义的资源名称属性。资源中引用资源，引用格式：@[package:]type:name，@表示对资源的引用，package 是包名称，如果在相同的包，package 则可以省略。android.jar 文件是 Android 程序所能引用的函数库文件，Android通过平台所支持API都包含在这个文件中。assets目录用来存放原始格式的文件，例如音频文件、视频文件等二进制格式文件。此目录中的资源不能被R.java文件索引，所以只能以字节流的形式读取，一般情况下为空。

res目录是资源目录，有三个子目录用来保存Android程序所有资源。drawable目录用来保存图像文件，layout目录用来保存与用户界面相关的布局文件，valuse目录保存文件颜色、风格、主题和字符串等。在Hello Android工程中，ADT在drawable目录中自动引入了icon.png文件，作为HelloAndroid程序的图标文件；在layout目录生成了mail.xml文件，用于描述用户界面。

main.xml文件是界面布局文件，利用XML语言描述用户界面，界面布局的相关内容将在第5章用户界面设计中进行详细介绍。

在main.xml文件的代码中，第7行的代码说明在界面中使用TextView控件，TextView控件主要用来显示字符串文本。第10行代码说明TextView控件需要显示的字符串，非常明显，@string/hello是对资源的引用。

```
1. <? xml version="1.0" encoding="utf-8"? >
2. <LinearLayout xmlns:android="http://schemas.android.com/apk/res/android"
3.     android:orientation="vertical"
4.     android:layout_width="fill_parent"
5.     android:layout_height="fill_parent"
6.     >
7. <TextView
8.     android:layout_width="fill_parent"
9.     android:layout_height="wrap_content"
10.    android:text="@string/hello"
11.    />
12. </LinearLayout>
```

strings.xml文件的代码如下。

```
1. <? xml version="1.0" encoding="utf-8"? >
2. <resources>
3.     <string name="hello">Hello World, HelloAndroid! </string>
4.     <string name="app_name">HelloAndroid</string>
5. </resources>
```

通过对strings.xml文件第3行代码的分析，在TextView控件中显示的字符串应是"Hello World, HelloAndroid!"。如果读者修改strings.xml文件的第3行代码的内容，重新编译、运行后，模拟器中显示的结果也应该随之更改。

AndroidManifest.xml是XML格式的Android程序声明文件，包含了Android系统运行Android程序前所必须掌握的重要信息。这些信息包含应用程序名称、图标、包名称、模块组成、授权和SDK最低版本等，而且每个Android程序必须在根目录下包含一个Android

Manifest.xml 文件。

AndroidManifest.xml 文件的代码如下。

```xml
1. <?xml version="1.0" encoding="utf-8"?>
2. <manifest xmlns:android="http://schemas.android.com/apk/res/android"
3.     package="edu.xzceu.HelloAndroid"
4.     android:versionCode="1"
5.     android:versionName="1.0">
6.   <application android:icon="@drawable/icon"
7.       android:label="@string/app_name">
8.     <activity android:name=".HelloAndroid"
9.         android:label="@string/app_name">
10.       <intent-filter>
11.         <action android:name="android.intent.action.MAIN" />
12.         <category android:name="android.intent.category.LAUNCHER" />
13.       </intent-filter>
14.     </activity>
15.   </application>
16.   <uses-sdk android:minSdkVersion="3" />
17. </manifest>
```

AndroidManifest.xml 文件的根元素是 manifest，包含了 xmlns:android、package、android:versionCode 和 android:versionName 共 4 个属性。xmlns:android 定义了 Android 的命名空间，其值为 http://schemas.android.com/apk/res/android，package 定义了应用程序的包名称。android:versionCode 定义了应用程序的版本号，是一个整数值，数值越大说明版本越新，但仅在程序内部使用，并不提供给应用程序的使用者，android:versionName 定义了应用程序的版本名称，是一个字符串，仅限于为用户提供一个版本标识。manifest 元素仅能包含一个 application 元素，application 元素中能够声明 Android 程序中最重要的四个组成部分，包括 Activity、Service、BroadcastReceiver 和 ContentProvider，所定义的属性将影响所有组成部分。第 6 行属性 android:icon 定义了 Android 应用程序的图标，其中@drawable/icon 是一种资源引用方式，表示资源类型是图像，资源名称为 icon，对应的资源文件为 res/drawable 目录下的 icon.png。第 7 行属性 android:label 则定义了 Android 应用程序的标签名称。activity 元素是对 Activity 子类的声明，必须在 AndroidManifest.xml 文件中声明的 Activity 才能在用户界面中显示。第 8 行属性 android:name 定义了实现 Activity 类的名称，可以是完整的类名称，也可以是简化后的类名称。第 9 行属性 android:label 则定义了 Activity 的标签名称，标签名称将在用户界面的 Activity 上部显示，@string/app_name 同样属于资源引用，表示资源类型是字符串，资源名称为 app_name，资源保存在 res/values 目录下的 strings.xml 文件中。intent-filter 中声明了两个子元素 action 和 category，intent-filter 使 HelloAndroid 程序在启动时，将.HelloAndroid 这个 Activity 作为默认启动模块。

可视化编辑器，双击 AndroidManifest.xml 文件，直接进入可视化编辑器，用户可以直接编辑 Android 工程的应用程序名称、包名称、图标、标签和许可等相关属性,如图 3-12 所示。

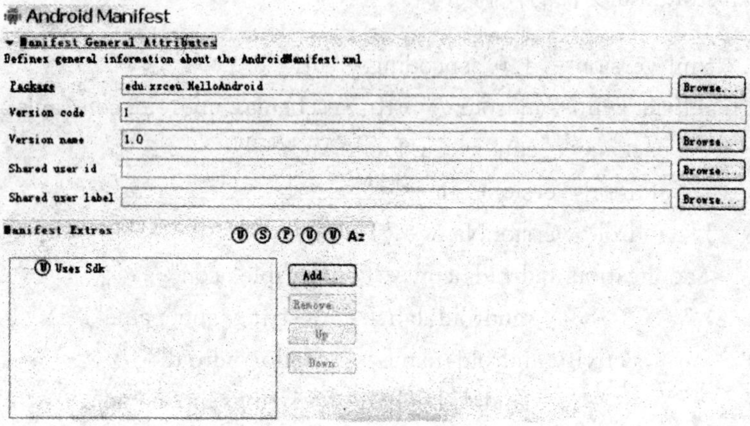

图 3-12 可视化编辑器

default.properties 文件记录 Android 工程的相关设置,该文件不能手动修改,需右键单击工程名称,选择"Properties"进行修改。在 default.properties 文件中只有第 12 行是有效代码,说明 Android 程序的编译目标。

```
1. # This file is automatically generated by Android Tools.
2. # Do not modify this file -- YOUR CHANGES WILL BE ERASED!
3. #
4. # This file must be checked in Version Control Systems.
5. #
6. # To customize properties used by the Ant build system use,
7. # "build.properties", and override values to adapt the script to your
8. # project structure.
9. # Project target.
10. target=android-3
```

3.3 命令行创建程序

命令行工具保存在<Android SDK>/tools 目录下,利用命令行工具开发 Android 程序的步骤如下:使用 android.bat 建立 HelloCommondline 工程所需的目录和文件,接着使用 Apache Ant 对 HelloCommondline 工程进行编译和 apk 打包,最后通过 adb.exe 将 HelloCommondline 工程上传到 Android 模拟器中。

使用 android.bat 建立 HelloCommondline 工程所需的目录和文件,android.bat 是一个批处理文件,可以用来建立和更新 Android 工程,管理 AVD,创建 Android 工程所需要的目录结构和文件。android.bat 建立和更新 Android 工程的命令和参数说明如表 3-1 所示。

表 3-1　Android.bat 建立和更新 Android 工程的命令和参数说明

命令	参数	说明	备注
android create project	-k <package>	包名称	必备参数
	-n <name>	工程名称	
	-a <activity>	Activity 名称	
	-t <target>	新工程的编译目标	必备参数
	-p <path>	新工程的保存路径	必备参数
android update project	-t <targe>	设定工程的编译目标	必备参数
	-p <path>	工程的保存路径	必备参数
	-n <name>	工程名称	

建立过程：开始→运行→CMD 启动 CMD 并进入<Android SDK>/tools 目录，输入命令

> android create project-n HelloCommandline-k edu.xzceu.HelloCommandline-a HelloCommandline-t 2-p e:\Android\workplace\HelloCommandline

或者

> android create project - -name HelloCommandline - -package edu.xzceu.HelloCommandline - -activity HelloCommandline - -target 2 - -path e:\Android\workplace\HelloCommandline

新工程的名称为 HelloCommandline，包名称为 edu.xzceu.HelloCommandline，Activity 名称是 HelloCommandline，编译目标的 ID 为 2，新工程的保存路径是 E:\Android\workplace\HelloCommandline。运行结果如图 3-13 所示。

图 3-13　工程运行的结果

仔细观察 android.bat 建立的目录和文件，发现其中一些在 Eclipse 开发环境中从未出现过的目录和文件，例如 build.xml、local.properties 和 tests 目录。这些新目录和文件的出现，主要是为了在构建 Android 程序时使用 Apache Ant。Apache Ant 是一个将软件编译、测试、部署等步骤联系在一起的自动化工具，多用于 Java 环境中的软件开发，若在构建 Android 程序时使用 Apache Ant，可以简化程序的编译和 apk 的打包过程。

HelloCommondline 工程文件和目录如表 3-2 所示。

表 3-2 HelloCommondline 工程文件和目录

文 件	说 明
AndroidManifest.xml	应用程序声明文件
build.xml	Ant 的构建文件
default.properties	保存编译目标，由 Android 工具自动建立，不可手工修改
build.properties	保存自定义的编译属性
local.properties	保存 Android SDK 的路径，仅供 Ant 使用
edu\xzceu\HelloCommandline/HelloCommandline.java	Activity 文件
bin/	编译脚本输出目录
gen/	保存 Ant 自动生成文件的目录，例如 R.java
libs/	私有函数库目录
res/	资源目录
src/	源代码目录
tests/	测试目录

在 HelloCommondline 工程文件和目录列表，libs 目录用来保存私有的函数库文件，在工程创建初期是空文件夹；tests 目录用于测试用途，在工程创建初期，文件夹的内容是 HelloCommandline 工程所有文件和目录的一个完成拷贝；local.properties 文件是保存 Android SDK 的路径的文件，由 Android 工具自动建立，不允许进行手工修改；local.properties 文件的主要用途是供 Apache Ant 寻找 Android SDK 的保存路径，第 10 行说明了 Android SDK 的路径是 E:\Android\android-sdk-windows-1.5_r1。

local.properties 文件的代码如下。

```
1. # This file is automatically generated by Android Tools.
2. # Do not modify this file -- YOUR CHANGES WILL BE ERASED!
3. #
4. # This file must *NOT* be checked in Version Control Systems,
5. # as it contains information specific to your local configuration.
6.
7. # location of the SDK. This is only used by Ant
8. # For customization when using a Version Control System, please read the
```

```
 9. # header note.
10. dk-location=E:\\Android\\android-sdk-windows-1.5_r1
```

build.properties 是保存自定义的编译属性的文件,能够修改应用程序的包名称、源代码目录和编译脚本输出目录等 Apache Ant 编译属性。在工程建立初期,build.properties 不包含任何有效代码,用户可以手工修改文件内容。如果需要修改应用程序的包名称,可以取消第 8 行的注释符号#,并将 com.example.myproject 替换为正确的包名称。修改源代码目录和编译脚本输出目录的位置,分别在 build.properties 文件的第 11 行和第 14 行。

build.properties 文件的代码如下:

```
 1. # This file is used to override default values used by the Ant build system.
 2. # This file must be checked in Version Control Systems, as it is
 3. # integral to the build system of your project.
 4. # The name of your application package as defined in the manifest.
 5. # Used by the 'uninstall' rule.
 6. # application-package=com.example.myproject
 7. # The name of the source folder.
 8. # source-folder=src
 9. # The name of the output folder.
10. # out-folder=bin
```

build.xml 是 Apache Ant 的构建文件,为编译 Android 程序提供基础信息,去除注释后的 build.xml 文件代码如下:

```
 1. <?xml version="1.0" encoding="UTF-8"?>
 2. <project name="HelloCommandline" default="help">
 3.     <property file="local.properties"/>
 4.     <property file="build.properties"/>
 5.     <property file="default.properties"/>
 6.     <path id="android.antlibs">
 7.         <pathelement path="${sdk-location}/tools/lib/anttasks.jar" />
 8.         <pathelement path="${sdk-location}/tools/lib/sdklib.jar" />
 9.         <pathelement path="${sdk-location}/tools/lib/androidprefs.jar" />
10.         <pathelement path="${sdk-location}/tools/lib/apkbuilder.jar" />
11.         <pathelement path="${sdk-location}/tools/lib/jarutils.jar" />
12.     </path>
13.     <taskdef name="setup"
14.         classname="com.android.ant.SetupTask"
15.         classpathref="android.antlibs"/>
16.     <setup />
17. </project>
```

build.xml 文件的第 4 行至第 6 行代码分别说明三个属性文件的名称。第 8 行至 14 行代码说明了构建过程中使用到的库文件的路径。

使用 Apache Ant 对 HelloCommondline 工程进行编译和 apk 打包,下载地址为:http://ant.apache.org/bindownload.cgi。网站提供 zip、tar.gz 和 tar.bz2 三种格式下载,Windows 系统用户推荐下载 zip 格式的二进制包。这里下载的 Apache Ant 压缩包为 apache-ant-1.7.1-bin.zip,版本号为 1.7.1,并将其解压缩在 E:\Android 目录下。

在 Windows 系统中添加新的环境变量,Apache 才能正常运行。修改位置:"我的电脑"→"属性"→"高级"→"环境变量"→"系统变量",新增的系统环境变量,表 3-3 所示。

表 3-3 系统环境变量表

变量名	变量值	备注
JAVA_HOME	C:\Program Files\Java\jdk_12	新增变量
ANT_HOME	E:\Android\apache-ant-	新增变量
ANDROID_HOME	E:\Android\android-sdk-windows-1.5_r1	新增变量
CLASSPATH	$JAVA_HOME/jre/lib;$JAVA_HOME/lib;$JAVA_HOME/lib/tools.jar	新增变量
Path	;%ANT_HOME%\bin;%JAVA_HOME%/bin;%ANDROID_HOME%\tools	已有变量

JAVA_HOME 是 JDK 的安装目录,根据 JDK 实际安装位置进行修改。ANT_HOME 是 Apache Ant 的安装目录,根据 Apache Ant 实际安装位置进行修改。ANDROID_HOME 是 Android SDK 的安装目录,根据实际安装位置进行修改。

CLASSPATH 是需要使用的库文件的位置,Path 是可执行文件的搜索路径,将<Apache Ant>/bin、<JDK>/bin 和<Android SDK>/tools 三个目录追加到原有的 Path 变量值中,目录之间使用分号分隔。

判断环境变量的正确性。在 CMD 中运行输入 ant 命令,通过命令的输出信息判断环境变量是否设置正确,如果输出的提示包含"Unable to locate tools.jar. Expected to find it in…",则表明设置环境变量不正确。如果环境变量设置正确,ant 命令的输出结果如下图 3-14。

图 3-14 ant 命令输出结果

数字签名机制。在 Android 平台上开发的所有应用程序都必须进行数字签名后,才能安装到模拟器或手机上,否则,将返回错误提示:Failure [INSTALL_PARSE_FAILED_NO_CERTIFICATERS]。

在 Eclipse 开发环境中,ADT 在将 Android 程序安装到模拟器前,已经利用内置的 debug key 为 apk 文件自动做了数字签名,这使用户无需自己生产数字签名的私钥,而能够利于 debug key 快速完成程序调试,但有一点需要注意,如果用户希望正式发布自己的应用程序,则不能使用 debug key,必须使用私有密钥对 Android 程序进行数字签名。

Apache Ant 构建 Android 应用程序支持 Debug 和 Release 两种模式:Debug 模式供调试使用,用于快速测试开发的应用程序,可以自动使用 debug key 完成数字签名;Release 模式是正式发布应用程序时使用的构建模式,生成没有数字签名的 apk 文件。

Debug 模式对 HelloCommandline 工程进行编译,生成具有 debug key 的 apk 打包文件。使用 CMD,在工程的根目录下,输入 ant debug,结果显示如图 3-15。

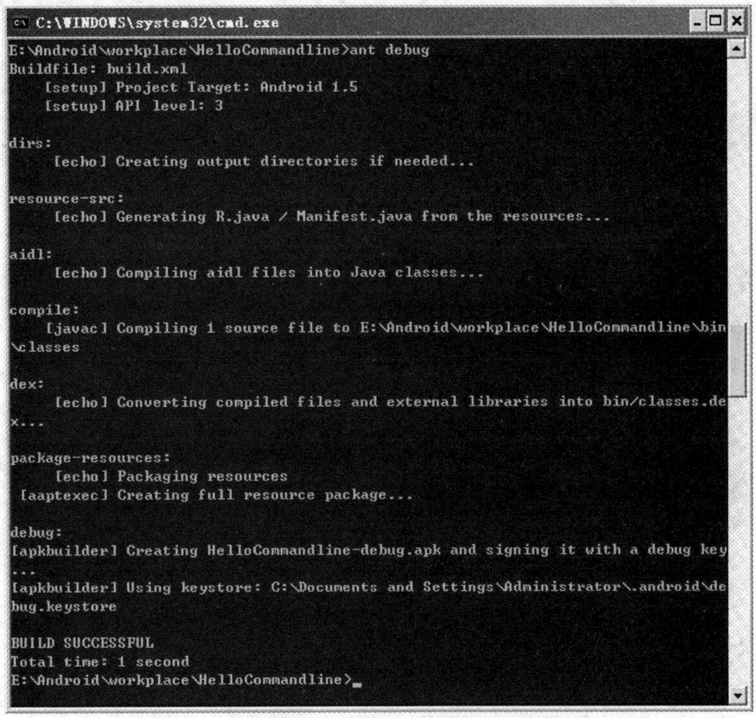

图 3-15 ant debug 输出结果

命令运行后,Apache Ant 在 bin 目录中生成打包文件 HelloCommandline-debugapk。如果需要使用 Release 模式,则需在 CMD 中输入 ant release,运行后会在 bin 目录中生成打包文件 HelloCommandline-unsignedapk。

apk 文件是 Android 系统的安装程序,上传到 Android 模拟器或 Android 手机后可以进行安装。apk 文件本身是一个 zip 压缩文件,能够使用 WinRAR、UnZip 等软件直接打开,下图 3-16 是使用 WinRAR 打开的 HelloCommandline-debug.apk 文件。

res 目录用来存放资源文件,AndroidManifest.xml 是 Android 声明文件,classes.dex 是 Dalvik 虚拟机的可执行程序,resources.arsc 是编译后的二进制资源文件。

使用 adb.exe 将 HelloCommondline 工程上传到 Android 模拟器中,在启动模拟器时,需指定所使用的 AVD。可以使用 android list avds 命令进行查询 AVD。在这里,建立两个 AVD,Android1.1 和 Android1.5,其中 1.1 版只是用于区别 1.5 版。这里使用 Android1.5 启动模拟器,在 CMD 中输入命令 emulator-avd Android1.5,如图 3-17 所示。

图 3-16　HelloCommandline-debug.apk 文件

图 3-17　emulator-avd Android1.5 运行结果

　　Android 模拟器正常启动后，利用 adb.exe 工具能够把 HelloCommandline－debugapk 文件上传到模拟器中。adb.exe 工具除了能够在 Android 模拟器中上传和下载文件，还能够管理模拟器状态，是调试程序时不可缺少的工具。在 CMD 中，进入＜HelloCommandline＞/bin 目录，输入命令：adb install HelloCommandline-debugapk，完成 apk 程序上传到模拟器的过程。如果上传成功，结果如图 3-18 所示。

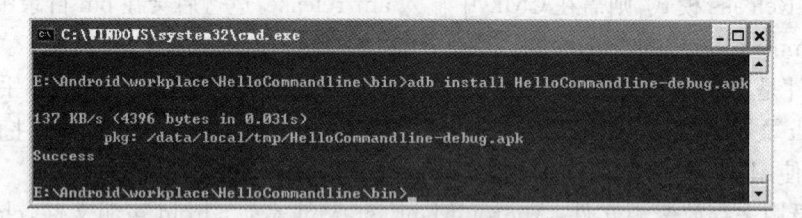

图 3-18　apk 上传成功

　　apk 文件上传后，需手工启动 HelloCommandline 程序。单击模拟器界面左下角上刚安装的 HelloCommandline 程序图标，即可手工启动。如果在模拟器中找不到新安装的程序，尝试重新启动 Android 模拟器。Android 的包管理器经常仅在模拟器启动时候检查应用程序的 AndroidManifest.xml 文件，这就导致部分上传的 Android 应用程序不能立即启动，如图 3-19 所示。

图 3‑19　HelloCommandline 应用程序

最后一步是修改 HelloCommandline 工程代码后,需要使用 Apache Ant 重新编译和打包应用程序,并将新生成的 apk 文件上传到 Android 模拟器中。如果新程序的包名称没有改变,则在使用 adb.exe 上传 apk 文件到模拟器时,会出现如下图 3‑20 的错误提示,此时,需要在模拟器中先删除原有 apk 文件,再使用 adb.exe 工具上传新的 apk 文件。

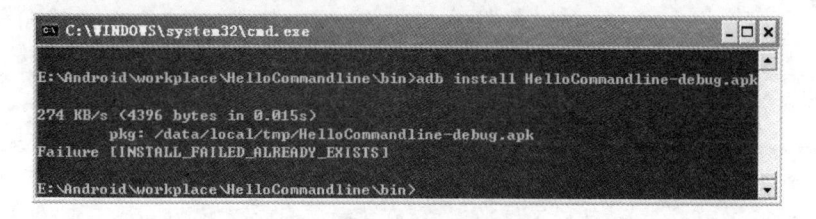

图 3‑20　错误提示示例

如果要删除 apk 文件,可以使用 adb uninstall ＜包名称＞的方法。例如删除 HelloCommandline 工程的 apk 文件,则可在 CMD 中输入命令 adb uninstall edu.xzceu. HelloCommandline,提示"Success"则表示成功删除。使用 adb shell rm /data/app/＜包名称＞apk 的方法,同样以删除 HelloCommandline 工程的 apk 文件为例,在 CMD 中输入下面的命令,没有任何提示则表示删除成功。

```
adb shell rm /data/app/edu.xzceu.HelloCommandlineapk
```

如果仅有一个 Android 模拟器在运行,用户可以一条命令完成 Android 工程编译、apk 打包和上传过程。启动 CMD,进入 HelloCommandline 工程的根目录下输入 adb install,adb. exe 将自动构建工程,并使用 debug key 对工程进行签名,之后将 apk 文件上传到 Android 模拟器中。如果同时有两个或两个以上的 Android 模拟器存在,这种方法将会失败,因为 adb. exe 不能够确定应该将 apk 文件上传到哪一个 Android 模拟器中。多次使用这种方法时,同样需要先删除模拟器中已有的 apk 文件。

习题与思考题

1. 简述 R.java 和 AndroidManefiest.xml 文件的用途。
2. 尝试建立一个 Android 2.1 版本的 AVD,AVD 的名称为 MyAVD_2.1。
3. 使用 Eclipse 建立名为 MyAndroid 的工程,包名称为 edu.xzceu.MyAndroid,使用第 2 题中建立的 AVD,程序运行时显示 Hello MyAndroid。
4. 尝试使用命令行方式建立一个 Android 应用程序,并完成 apk 打包和程序安装过程。

第 4 章 Android 生命周期

☆4.1 程序生命周期
☆4.2 Android 组件
☆4.3 Activity 生命周期
☆4.4 程序调试

本章学习目标：了解 Android 系统的进程优先级的变化方式，了解 Android 系统的四大基本组件，了解 Activity 的生命周期中各状态的变化关系；掌握 Activity 事件回调函数的作用和调用顺序，掌握 Android 应用程序的调试方法和工具。

4.1 程序生命周期

程序的生命周期是进程在 Android 系统中从启动到终止的所有阶段，也就是 Android 程序启动到停止的全过程。程序的生命周期是由 Android 系统进行调度和控制的。

Android 系统中的进程按优先级由高到低顺序包括：前台进程、可见进程、服务进程、后台进程、空进程等，如图 4-1 所示。

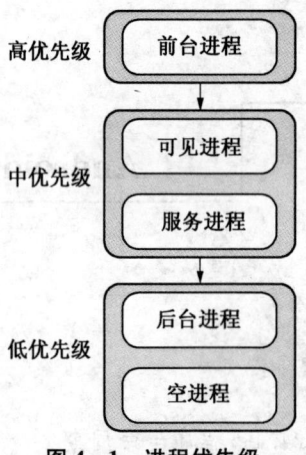

图 4-1 进程优先级

（1）前台进程。前台进程是 Android 系统中最重要的进程，是与用户正在交互的进程，包含以下四种情况：进程中的 Activity 正在与用户进行交互；进程服务被 Activity 调用，而且这个 Activity 正在与用户进行交互；进程服务正在执行声明周期中的回调函数，如 onCreate()、onStart()或 onDestroy()；进程的 Broadcast Receiver 正在执行 onReceive()函数。Android 系统在多个前台进程同时运行时，可能会出现资源不足的情况，此时会清除部分前台进程，保证主要的用户界面能够及时响应。

（2）可见进程。可见进程指部分程序界面能够被用户看见，却不在前台与用户交互，不响应界面事件的进程。如果一个进程包含服务，且这个服务正在被用户可见的 Activity 调用，此进程同样被视为可见进程。Android 系统一般存在少量的可见进程，只有在特殊的情况下，Android 系统才会为保证前台进程的资源而清除可见进程。

（3）服务进程。服务进程是指包含已启动服务的进程，没有用户界面，在后台长期运行。除非 Android 系统不能保证前台进程或可视进程所必要的资源，否则不强行清除服务进程。

（4）后台进程。后台进程是指不包含任何已经启动的服务，而且没有任何用户可见的 Activity 的进程，Android 系统中一般存在数量较多的后台进程，在系统资源紧张时，系统将优先清除用户较长时间没有用到的后台进程。

（5）空进程。空进程是不包含任何活跃组件的进程。空进程在系统资源紧张时会被首先清除，但为了提高 Android 系统应用程序的启动速度，Android 系统会将空进程保存在系统内存中,在用户重新启动该程序时,空进程会被重新使用。除了以上的优先级外,以下两方面也决定他们的优先级,进程的优先级取决于所有组件中的优先级最高的部分,进程的优先级会根据与其他进程的依赖关系而变化。

4.2 Android 组件

Android 组件是可以调用的基本功能模块,Android 应用程序就是由组件组成的。Android 系统有四个重要的组件,分别是 Activity、Service、Broadcast Receiver 和 Content Provider。

(1) Activity。Activity 是 Android 程序的呈现层,显示可视化的用户界面,并接收与用户交互所产生的界面事件。Android 应用程序可以包含一个或多个 Activity,一般在程序启动后会呈现一个 Activity,用于提示用户程序已经正常启动,在界面上的表现形式有全屏窗体,非全屏悬浮窗体,对话框等。

(2) Service。Service 用于没有用户界面,但需要长时间在后台运行的应用。

(3) Broadcast Receiver。Broadcast Receiver 是用来接受并响应广播消息的组件。不包含任何用户界面,可以通过启动 Activity 或者 Notification 通知用户接收到重要信息。Notification 能够通过多种方法提示用户,包括闪动背景灯、振动设备、发出声音或在状态栏上放置一个持久的图标等。

(4) Content Provider。Content Provider 是 Android 系统提供的一种标准的共享数据的机制。应用程序可以通过 Content Provider 访问其他应用程序的私有数据。私有数据可以是存储在文件系统中的文件,也可以是 SQLite 中的数据库。Android 系统内部也提供一些内置的 Content Provider,能够为应用程序提供重要的数据信息。

所有 Android 组件都具有自己的生命周期,是组件从建立到销毁的整个过程。在生命周期中,组件会在可见、不可见、活动、非活动等状态中不断变化。

4.3 Activity 生命周期

Activity 生命周期指 Activity 从启动到销毁的整个过程,如图 4-2 所示。Activity 表现为四种状态,分别是活动状态、暂停状态、停止状态和非活动状态。活动状态时 Activity 在用户界面中处于最上层,完全能被用户看到,能够与用户进行交互。暂停状态时 Activity 在界面上被部分遮挡,该 Activity 不再处于用户界面的最上层,且不能够与用户进行交互。停止状态时 Activity 在界面上完全不能被用户看到,也就是说这个 Activity 被其他 Activity 全部遮挡。

图 4-2 Activity 的四种状态的变换关系

不在以上三种状态中的Activity则处于非活动状态。如图4-3所示。

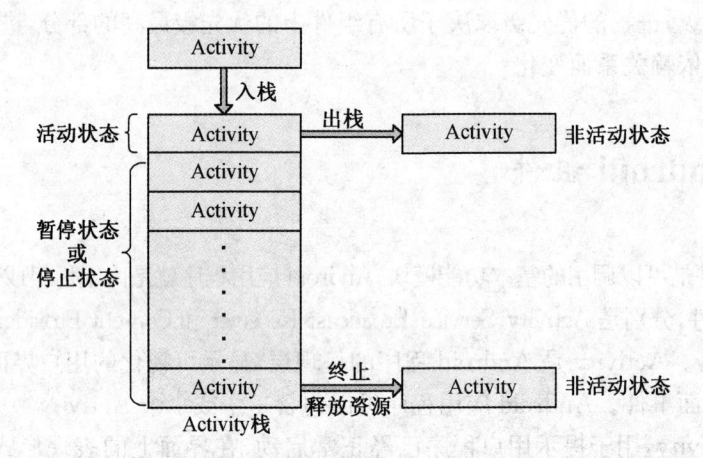

图4-3 Activity栈,遵循"后进先出"的规则

对应于Activity生命周期,有相应的事情回调函数,事件的回调函数如表4-1、4-2所示。回调函数的调用顺序,如图4-4所示。

表4-1 Activity生命周期的事件回调函数

函数	是否可终止	说明
onCreate()	否	Activity启动后第一个被调用的函数,常用来进行Activity的初始化,例如创建View、绑定数据或恢复信息等。
onStart()	否	当Activity显示在屏幕上时,该函数被调用。
onRestart()	否	当Activity从停止状态进入活动状态前,调用该函数。
onResume()	否	当Activity能够与用户交互,接受用户输入时,该函数被调用。此时的Activity位于Activity栈的栈顶。
onPause()	是	当Activity进入暂停状态时,该函数被调用。一般用来保存持久的数据或释放占用的资源。
onStop()	是	当Activity进入停止状态时,该函数被调用。
onDestroy()	是	在Activity被终止前,即进入非活动状态前,该函数被调用。

表4-2 Activity状态保存/恢复的事件回调函数

函数	是否可终止	说明
onSaveInstanceState()	否	Android系统因资源不足终止Activity前调用该函数,用以保存Activity的状态信息,供onRestoreInstanceState()或onCreate()恢复之用。
onRestoreInstanceState()	否	恢复onSaveInstanceState()保存的Activity状态信息,在onStart()和onResume()之间被调用。

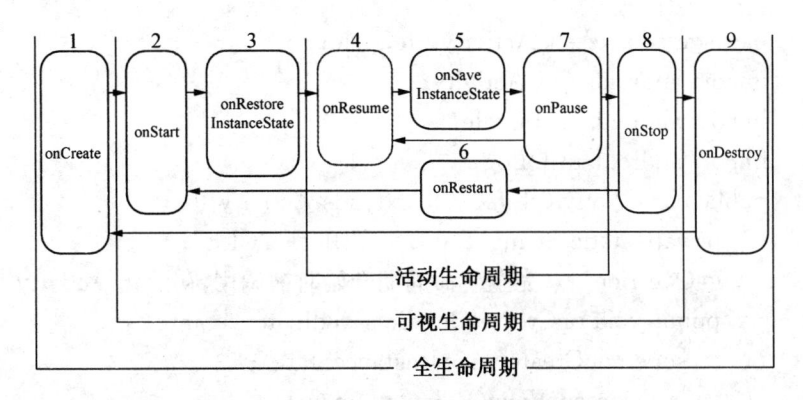

图 4-4　Activity 事件回调函数的调用顺序

Activity 的生命周期可分为全生命周期、可视生命周期和活动生命周期,如图 4-4 所示。每种生命周期中包含不同的事件回调函数。

(1) 全生命周期。全生命周期是从 Activity 建立到销毁的全部过程,始于 onCreate(),结束于 onDestroy()。使用者通常在 onCreate()中初始化 Activity 所能使用的全局资源和状态,并在 onDestroy()中释放这些资源。在一些极端的情况下,Android 系统会不调用 onDestroy()函数,而直接终止进程。

(2) 可视生命周期。可视生命周期是 Activity 在界面上从可见到不可见的过程,开始于 onStart(),结束于 onStop()。onStart()一般用来初始化或启动与更新界面相关的资源,onStop()一般用来暂停或停止一切与更新用户界面相关的线程、计时器和服务,onRestart()函数在 onStart()前被调用,用来在 Activity 从不可见变为可见的过程中,进行一些特定的处理过程。onStart()和 onStop()会被多次调用,onStart()和 onStop()也经常被用来注册和注销 Broadcast Receiver。

(3) 活动生命周期。活动生命周期是 Activity 在屏幕的最上层,并能够与用户交互的阶段,开始于 onResume(),结束于 onPause()。在 Activity 的状态变换过程中 onResume()和 onPause()经常被调用,因此这两个函数中应使用更为简单、高效的代码。onPause()是第一个被标识为"可终止"的函数。在 onPause()返回后,onStop()和 onDestroy()随时能被 Android 系统终止,onPause()常用来保存持久数据,如界面上的用户的输入信息等。

onPause()和 onSaveInstanceState()这两个函数都可以用来保存界面的用户输入数据,他们的区别在于,onPause()一般用于保存持久性数据,并将数据保存在存储设备上的文件系统或数据库系统中的,onSaveInstanceState()主要用来保存动态的状态信息,信息一般保存在 Bundle 中,Bundle 是能够保存多种格式数据的对象。onSaveInstanceState()将数据保存在 Bundle 中,系统在调用 onRestoreInstanceState()和 onCreate()时,会同样利用 Bundle 将数据传递给函数。

举例:建立一个新的 Android 工程,工程名称:ActivityLifeCycle,包名称:edu. xzceu. ActivityLifeCycle,Activity 名称:ActivityLifeCycle。

ActivityLifeCycle.java 文件的代码如下。

```
1.    package edu.xzceu.ActivityLifeCycle;
2.    import android.app.Activity;
3.    import android.os.Bundle;
4.    import android.util.Log;
5.    public class ActivityLifeCycle extends Activity {
6.        private static String TAG = "LIFTCYCLE";
7.        @Override    //完全生命周期开始时被调用,初始化 Activity
8.        public void onCreate(Bundle savedInstanceState) {
9.            super.onCreate(savedInstanceState);
10.           setContentView(R.layout.main);
11.           Log.i(TAG, "(1) onCreate()");
12.       }
13.       @Override    //可视生命周期开始时被调用,对用户界面进行必要的更改
14.       public void onStart() {
15.           super.onStart();
16.           Log.i(TAG, "(2) onStart()");
17.       }
18.       @Override    //在 onStart()后被调用,用于恢复 onSaveInstanceState()保存的用户界面信息
19.       public void onRestoreInstanceState(Bundle savedInstanceState) {
20.           super.onRestoreInstanceState(savedInstanceState);
21.           Log.i(TAG, "(3) onRestoreInstanceState()");
22.       }
23.
24.       @Override    //在活动生命周期开始时被调用,恢复被 onPause()停止的用于界面更新的资源
25.       public void onResume() {
26.           super.onResume();
27.           Log.i(TAG, "(4) onResume()");
28.       }
29.
30.       @Override    // 在 onResume()后被调用,保存界面信息
31.       public void onSaveInstanceState(Bundle savedInstanceState) {
32.           super.onSaveInstanceState(savedInstanceState);
33.           Log.i(TAG, "(5) onSaveInstanceState()");
34.       }
35.
36.       @Override    //在重新进入可视生命周期前被调用,载入界面所需要的更改信息
37.       public void onRestart() {
```

```
38.        super.onRestart();
39.        Log.i(TAG, "(6) onRestart()");
40.    }
41.
42.    @Override   //在活动生命周期结束时被调用,用来保存持久的数据或
                    释放占用的资源
43.    public void onPause() {
44.        super.onPause();
45.        Log.i(TAG, "(7) onPause()");、
46.    }
47.
48.    @Override  //在可视生命周期结束时被调用,一般用来保存持久的数据或
                   释放占用的资源
49.    public void onStop() {
50.      super.onStop();
51.      Log.i(TAG, "(8) onStop()");
52.    }
53.
54.    @Override  //在完全生命周期结束时被调用,释放资源,包括线程、数据连接等
55.    public void onDestroy() {
56.        super.onDestroy();
57.        Log.i(TAG, "(9) onDestroy()");
58.    }
59. }
```

上面的程序主要通过在生命周期函数中添加"日志点"的方法进行调试,程序的运行结果将会显示在 LogCat 中。为了显示结果易于观察和分析,在 LogCat 设置过滤器 LifeCycleFilter,过滤方法选择 by Log Tag,过滤关键字为 LIFTCYCLE。

在全生命周期中启动和关闭 ActivityLifeCycle 的 LogCat 输出的方法:先启动 ActivityLifeCycle,然后按下模拟器上的"返回键",关闭 ActivityLifeCycle 的 LogCat 输出结果,如图 4-5 所示。

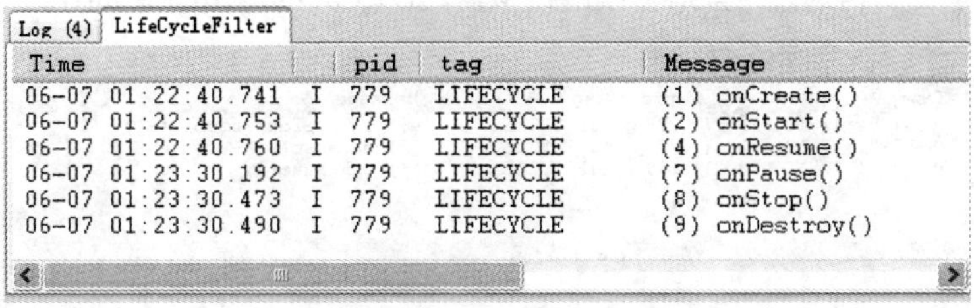

图 4-5 全生命周期的 LogCat 输出

函数的调用顺序:onCreate()→onStart()→onResume()→onPause()→onStop()→onDestroy()。调用 onCreate()函数分配资源,调用 onStart()将 Activity 显示在屏幕上,调用

onResume()获取屏幕焦点，调用 onPause()、onStop()和 onDestroy()，释放资源并销毁进程。

可视生命周期的状态转换：启动 ActivityLifeCycle，按“呼出/接听键”启动内置的拨号程序，再通过“返回键”退出拨号程序，ActivityLifeCycle 重新显示在屏幕中。可视生命周期的 LogCat 输出结果如图 4-6 所示。

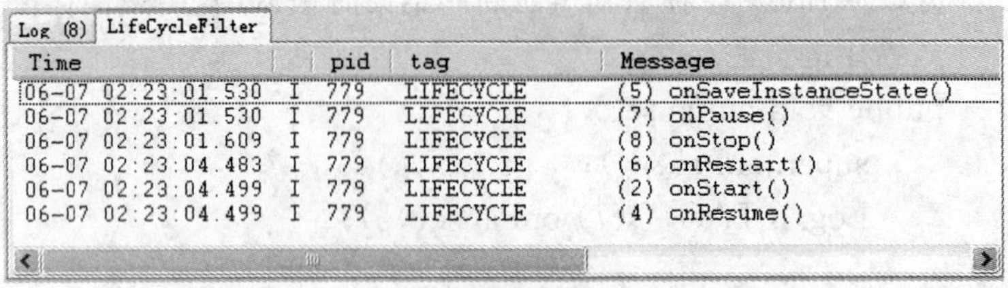

Log (8) | LifeCycleFilter

Time		pid	tag	Message
06-07 02:23:01.530	I	779	LIFECYCLE	(5) onSaveInstanceState()
06-07 02:23:01.530	I	779	LIFECYCLE	(7) onPause()
06-07 02:23:01.609	I	779	LIFECYCLE	(8) onStop()
06-07 02:23:04.483	I	779	LIFECYCLE	(6) onRestart()
06-07 02:23:04.499	I	779	LIFECYCLE	(2) onStart()
06-07 02:23:04.499	I	779	LIFECYCLE	(4) onResume()

图 4-6　可视生命周期的 LogCat 输出

函数的调用顺序：onSaveInstanceState()→onPause()→onStop()→onRestart()→onStart()→onResume()，调用 onSaveInstanceState()函数保存 Activity 状态，调用 onPause()和 onStop()，停止对不可见 Activity 的更新，调用 onRestart()恢复界面上需要更新的信息，调用 onStart()和 onResume()重新显示 Activity，并接受用户交互。

开启 IDA 的可视生命周期：Dev Tools→Development Settings→Immediately destroy activities (IDA)下开启 IDA，如图 4-7 所示。

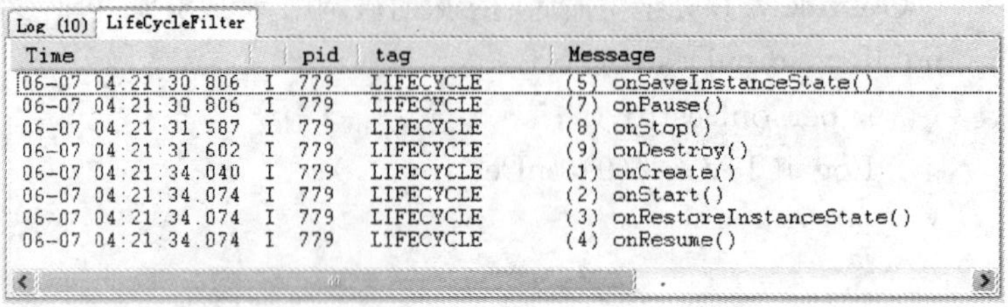

Log (10) | LifeCycleFilter

Time		pid	tag	Message
06-07 04:21:30.806	I	779	LIFECYCLE	(5) onSaveInstanceState()
06-07 04:21:30.806	I	779	LIFECYCLE	(7) onPause()
06-07 04:21:31.587	I	779	LIFECYCLE	(8) onStop()
06-07 04:21:31.602	I	779	LIFECYCLE	(9) onDestroy()
06-07 04:21:34.040	I	779	LIFECYCLE	(1) onCreate()
06-07 04:21:34.074	I	779	LIFECYCLE	(2) onStart()
06-07 04:21:34.074	I	779	LIFECYCLE	(3) onRestoreInstanceState()
06-07 04:21:34.074	I	779	LIFECYCLE	(4) onResume()

图 4-7　IDA 的可视生命周期的 LogCat 输出

开启 IDA 的可视生命周期的函数调用顺序：onSaveInstanceState() →onPause()→onStop()→onDestroy()→onCreate()→onStart()→onRestoreInstanceState()→onResume()，调用 onRestoreInstanceState()恢复 Activity 销毁前的状态，其他的函数调用顺序与程序启动过程的调用顺序相同。

在活动生命周期显示活动生命周期的 LogCat 输出：启动 ActivityLifeCycle，通过“挂断键”使模拟器进入休眠状态，再通过“挂断键”唤醒模拟器。LogCat 的输出结果如图 4-8 所示。

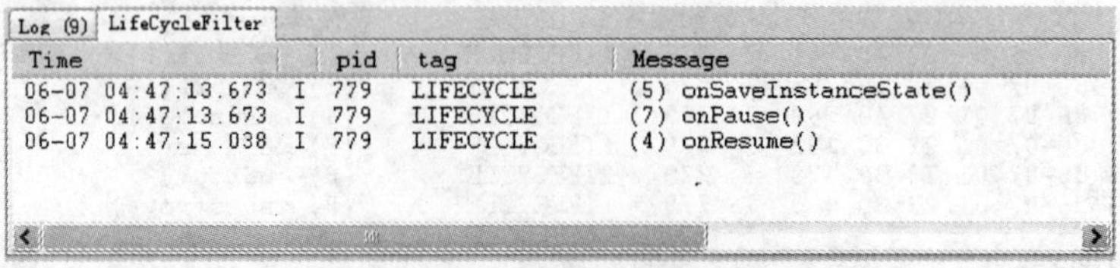

Log (9) | LifeCycleFilter

Time		pid	tag	Message
06-07 04:47:13.673	I	779	LIFECYCLE	(5) onSaveInstanceState()
06-07 04:47:13.673	I	779	LIFECYCLE	(7) onPause()
06-07 04:47:15.038	I	779	LIFECYCLE	(4) onResume()

图 4-8　活动生命周期的 LogCat 输出

函数调用顺序如下：onSaveInstanceState()→onPause()→onResume()，调用 onSaveInstanceState()保存 Activity 的状态，调用 onPause()停止与用户交互，调用 onResume()恢复与用户的交互。

4.4　程序调试

Android 系统提供了两种调试工具 LogCat 和 DevTools,用于定位、分析及修复程序中出现的错误。

4.4.1　LogCat

LogCat 是用来获取系统日志信息的工具,并可以显示在 Eclipse 集成开发环境中,能够捕获的信息包括 Dalvik 虚拟机产生的信息、进程信息、Activity Manager 信息、Packager Manager 信息、Homeloader 信息、Windows Manager 信息、Android 运行时信息和应用程序信息等。

打开方式:Window→Show View→Other,打开 Show View 的选择菜单,如图 4-9 所示。

图 4-9　打开 LogCat 界面

然后在 Andoird→LogCat 中选择 LogCat。LogCat 打开后,便显示在 Eclipse 的下方区域,如图 4-10 所示。

图 4-10　LogCat 显示区域界面

LogCat 的右上方的五个字母表示五种不同类型的日志信息，他们的级别依次增高：
［V］：详细（Verbose）信息
［D］：调试（Debug）信息
［I］：通告（Info）信息
［W］：警告（Warn）信息
［E］：错误（Error）信息

在 LogCat 中，用户可以通过五个字母图标选择显示的信息类型，级别高于所选类型的信息也会在 LogCat 中显示，但级别低于所选类型的信息则不会被显示。LogCat 提供了"过滤"功能，在右上角的"＋"号和"－"号，分别是添加和删除过滤器。用户可以根据日志信息的标签（Tag）、产生日志的进程编号（Pid）或信息等级（Level），对显示的日志内容进行过滤。

程序调试原理：引入 android.util.Log 包，使用 Log.v()、Log.d()、Log.i()、Log.w() 和 Log.e() 五个函数在程序中设置"日志点"，当程序运行到"日志点"时，应用程序的日志信息便被发送到 LogCat 中，判断"日志点"信息与预期的内容是否一致，进而判断程序是否存在错误。Log.v() 用来记录详细信息，Log.d() 用来记录调试信息，Log.i() 用来记录通告信息，Log.w() 用来记录警告信息，Log.e() 用来记录通讯错误信息。

下面的代码演示了 Log 类的具体使用方法。

```
1.  package edu.xzceu.LogCat;
2.  import android.app.Activity;
3.  import android.os.Bundle;
4.  import android.util.Log;
5.  public class LogCat extends Activity {
6.      final static String TAG = "LOGCAT";
7.      @Override
8.      public void onCreate(Bundle savedInstanceState) {
9.          super.onCreate(savedInstanceState);
10.         setContentView(R.layout.main);
11.         Log.v(TAG,"Verbose");
12.         Log.d(TAG,"Debug");
13.         Log.i(TAG,"Info");
14.         Log.w(TAG,"Warn");
15.         Log.e(TAG,"Error");
16.     }
17. }
```

程序第 5 行引入 android.util.Log 包，第 8 行的定义标签帮助用户在 LogCat 中找到目标程序生成的日志信息，同时也能够利用标签对日志进行过滤，第 13 行记录一个详细信息。Log.v() 函数的第一个参数是日志的标签，第二个参数是实际的信息内容，第 14 行到第 17 行分别产生了调试信息、通告信息、警告信息和错误信息。LogCat 工程的运行结果如图 4-11 所示，LogCat 对不同类型的信息使用了不同的颜色加以区别。

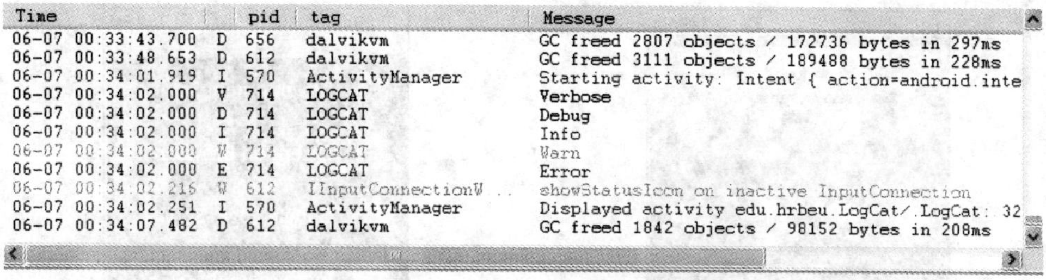

图 4-11 LogCat 工程的运行结果

添加过滤器：单击"＋"，填入过滤器的名称 LogcatFilter，设置过滤条件为"标签＝LOGCAT"，如图 4-12 所示。

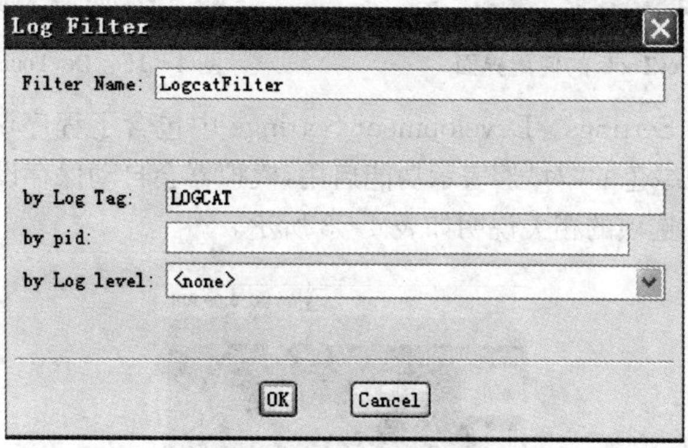

图 4-12 添加过滤器界面

LogCat 过滤后的输入结果，如图 4-13 所示。无论什么类型的日志信息，属于哪一个进程，只要标签为 LOGCAT，都将显示在 LogcatFilter 区域内。

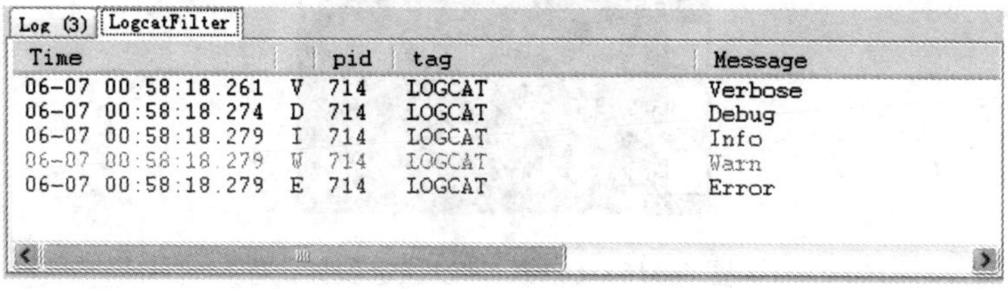

图 4-13 LogCat 过滤后结果

4.4.2 DevTools

DevTools 是用于调试和测试的工具，包括了一系列各种用途的小工具：Development Settings、Exception Browser、Google Login Service、Instrumentation、Media Scanner、Package Browser、Pointer Location、Raw Image Viewer、Running Processes 和 Terminal Emulator。DevTools 的使用界面如图 4-14 所示。

图 4-14　DevTools 的使用界面 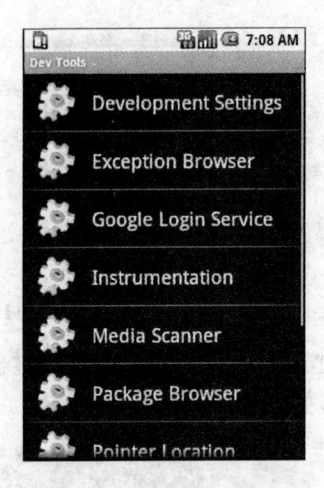　　　　　图 4-15　DevTools 设置界面

(1) Development Settings。Development Settings 中包含了程序调试的相关选项,单击功能前面选择框,出现绿色的"对号"表示功能启用,模拟器会自动保存设置,如图 4-16 所示。Development Settings 选项的相关说明如表 4-3 所示。

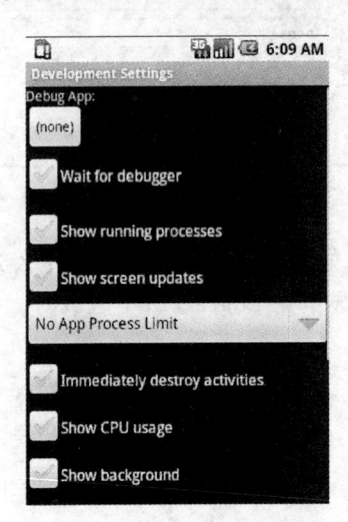

图 4-16　DevTools 设置选项界面

表 4-3　Development Settings 选项说明

选　项	说　明
Debug App	为 Wait for debugger 选项指定应用程序,如果不指定(选择 none),Wait for debugger 选项将适用于所有应用程序。Debug App 可以有效地防止 Android 程序长时间停留在断点而产生异常。
Wait for debugger	阻塞加载应用程序,直到关联到调试器(Debugger)。用于在 Activity 的 onCreate()函数进行断点调试。
Show running processs	在屏幕右上角显示运行中的进程。

续表

选项	说明
Show screen updates	选中该选项时，界面上任何被重绘的矩形区域会闪现粉红色，有利于发现界面中不必要的重绘区域。
No App Process limit	允许同时运行进程的数量上限。
Immediately destroy activites	Activity 进入停止状态后立即销毁，用于测试在函数 onSaveInstanceState()、onRestoreInstanceState()和 onCreate()中的代码。
Show CPU usage	在屏幕顶端显示 CPU 使用率，上层红线显示总的 CPU 使用率，下层绿线显示当前进程的 CPU 使用率。
Show background	应用程序没有 Activity 显示时，直接显示背景面板，一般这种情况仅在调试时出现。
Show Sleep state on LED	在休眠状态下开启 LED。
Windows Animation Scale	窗口动画显示比例。
Transition Animation	过度动画显示比例。
Light Hinting	提示模式设置。
Show GTalk service connection status	显示 GTalk 服务连接状态。

(2) Package Browser。Package Browser 是 Android 系统中的程序包查看工具，能够详细显示已经安装到 Android 系统中的程序信息，包括包名称、应用程序名称、图标、进程、用户 ID、版本、apk 文件保存位置和数据文件保存位置，进一步查看应用程序所包含 Activity、Service、Broadcast Receiver 和 Provider 的详细信息。如图 4－17 所示为查看 Activity LifeCycle 程序显示的相关信息。

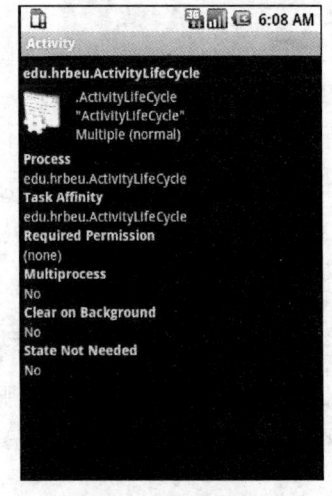

图 4－17 查看 ActivityLifeCycle 程序的相关信息

(3) Pointer Location。Pointer Location 是屏幕点位置查看工具,能够显示触摸点的 X 轴坐标和 Y 轴坐标。Pointer Location 的使用画面如图 4-18 所示。

(4) Running processes。Running processes 能够查看在 Android 系统中正在运行的进程,并能查看进程的详细信息,包括进程名称和进程所调用的程序包、Andoird 模拟器缺省情况下运行的进程和 com.android.phone 进程的详细信息,如图 4-19、4-20 所示。

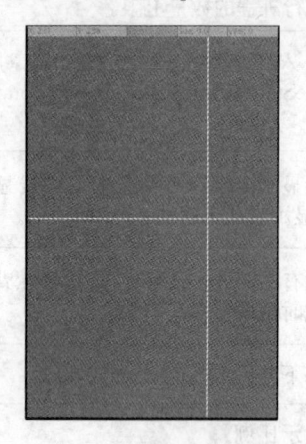

图 4-18　屏幕点位置查看工具　　　　图 4-19　进程查看器

图 4-20　进程的详细信息　　　　图 4-21　终端模拟器

(5) Terminal Emulator。Terminal Emulator 可以打开一个连接底层 Linux 系统的虚拟终端,但拥有的权限较低,且不支持提升权限的 su 命令,如果需要使用 root 权限的命令,可以使用 ADB 工具。如图 4-21 所示是 Terminal Emulator 运行时的画面,输入 ls 命令,显示出根目录下的所有文件夹。

习题与思考题

1. 简述 Android 系统前台进程、可见进程、服务进程、后台进程和空进程的优先级排序原因。
2. 简述 Android 系统的四种基本组件 Activity、Service、Broadcast Receiver 和 Content Provider 的用途。
3. 简述 Activity 生命周期的四种状态,以及状态之间的变换关系。
4. 简述 Activity 事件回调函数的作用和调用顺序。

第 5 章

Android 用户界面

☆5.1 用户界面基础
☆5.2 界面控件
☆5.3 界面布局
☆5.4 菜单
☆5.5 界面事件

本章学习目标：了解各种用户界面的控件的使用方法；掌握各种界面布局的特点和使用方法，掌握选项菜单、子菜单和快捷菜单的使用方法，掌握按键事件和触摸事件的处理方法。

5.1 用户界面基础

用户界面(User Interface,UI)是系统和用户之间进行信息交换的媒介，实现信息的内部形式与人类可以接受形式之间的转换。在计算机出现早期，批处理界面(1945—1968)和命令行界面(1969—1983)得到广泛的使用，目前流行采用图形方式与用户进行交互的图像用户界面(Graphical User Interface,GUI)。未来的用户界面将更多地运用虚拟现实技术，使用户能够摆脱键盘与鼠标的交互方式，而通过动作、语言，甚至是脑电波来控制计算机。

设计手机用户界面应解决的问题，需要界面设计与程序逻辑完全分离，这样不仅有利于他们的并行开发，而且在后期修改界面时，也不用再次修改程序的逻辑代码。根据不同型号手机的屏幕解析度、尺寸和纵横比，自动调整界面上部分控件的位置和尺寸，避免因为屏幕信息的变化而出现显示错误。能够合理利用较小的屏幕显示空间，构造出符合人机交互规律的用户界面，避免出现凌乱、拥挤的用户界面。Android 已经解决了前两个问题，使用 XML 文件描述用户界面作为资料文件；资源文件独立保存在资源文件夹中；对用户界面描述非常灵活，在不明确定义界面元素的位置和尺寸的情况下，仅声明界面元素的相对位置和粗略尺寸。

如图 5-1 所示为 Android 用户界面框架(Android UI Framework)，Android 用户界面框架采用 MVC(Model-View-Controller)模型，提供了处理用户输入的控制器(Controller)，显示用户界面和图像的视图(View)，以及保存数据和代码的模型(Model)。

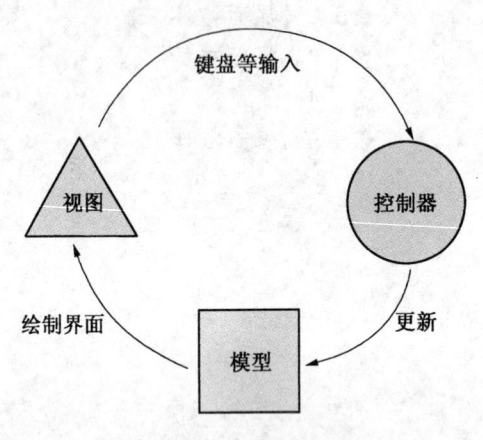

图 5-1 MVC 模型

MVC 模型中的控制器能够接受并响应程序的外部动作，如按键动作或触摸屏动作等。控制器使用队列处理外部动作，每个外部动作作为一个独立的事件被加入队列中，然后 Android 用户界面框架按照"先进先出"的规则从队列中获取事件，并将这个事件分配给所对应的事件处理函数。

如图 5-2 所示，Android 用户界面框架采用视图树(View Tree)模型，框架中的界面元素以一种树型结构组织在一起，称为视图树，Android 系统会依据视图树的结构从上至下绘制每一个界面元素。每个元素负责对自身的绘制,如果元素包含子元素,该元素会通知其下所有子元素进行绘制。

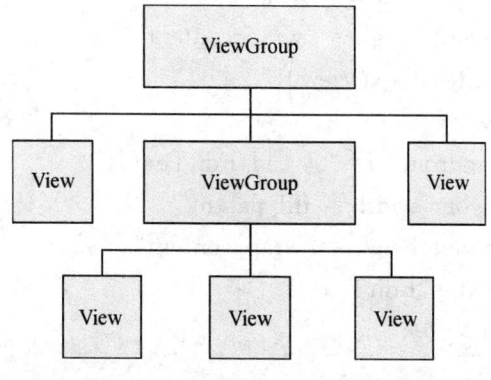

图 5-2 视图树(View Tree)模型

视图树由 View 和 ViewGroup 构成。View 是界面的最基本的可视单元,存储了屏幕上特定矩形区域内所显示内容的数据结构,并能够实现所占据区域的界面绘制、焦点变化、用户输入和界面事件处理等功能。View 也是一个重要的基类,所有在界面上的可见元素都是 View 的子类。ViewGroup 是一种能够承载含多个 View 的显示单元,它的功能一个是承载界面布局,另一个是承载具有原子特性的重构模块。

在单线程用户界面中,控制器从队列中获取事件和视图在屏幕上绘制用户界面,使用的都是同一个线程。特点:处理函数具有顺序性,能够降低应用程序的复杂程度,同时也能减低开发的难度,缺点:如果事件处理函数过于复杂,可能会导致用户界面失去响应。

5.2 界面控件

Android 系统的界面控件分为定制控件和系统控件。定制控件是用户独立开发的控件,或通过继承并修改系统控件后所产生的新控件,能够为用户提供特殊的功能或与众不同的显示需求方式。系统控件是 Android 系统提供给用户已经封装的界面控件。提供在应用程序开发过程中常见功能控件。系统控件更有利于帮助用户进行快速开发,同时能够使 Android 系统中应用程序的界面保持一致性,常见的系统控件包括 TextView、EditText、Button、ImageButton、Checkbox、RadioButton、Spinner、ListView 和 TabHost。

5.2.1 TextView 和 EditText

TextView 是一种用于显示字符串的控件,EditText 则是用来输入和编辑字符串的控件,EditText 是一个具有编辑功能的 TextView。建立一个"TextViewDemo"的程序,如图 5-3 所示,包含 TextView 和 EditText 两个控件,上方"用户名"部分使用的是 TextView,下方的文字输入框使用的是 EditText。

TextViewDemo 在 XML 文件中的代码如下。

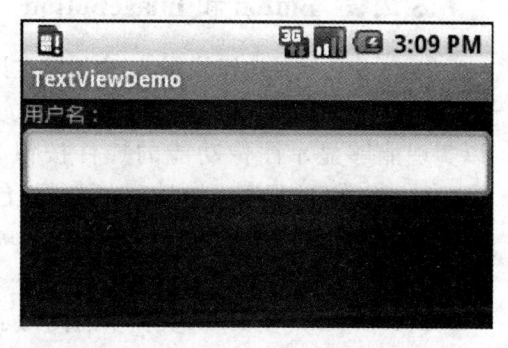

图 5-3 "TextViewDemo"视图效果

```
1. <TextView android:id="@+id/TextView01"
2.     android:layout_width="wrap_content"
3.     android:layout_height="wrap_content"
4.     android:text="TextView01" >
5. </TextView>
6. <EditText android:id="@+id/EditText01"
7.     android:layout_width="fill_parent"
8.     android:layout_height="wrap_content"
9.     android:text="EditText01" >
10. </EditText>
```

如以上代码所示,第 1 行 android:id 属性声明了 TextView 的 ID,这个 ID 主要用于在代码中引用这个 TextView 对象,"@+id/TextView01"表示所设置的 ID 值,@表示后面的字符串是 ID 资源,加号(+)表示需要建立新资源名称,并添加到 R.java 文件中,斜杠后面的字符串"TextView01"表示新资源的名称。如果资源不是新添加的,或属于 Android 框架的 ID 资源,则不需要使用加号(+),但必须添加 Android 包的命名空间,例如 android:id="@android:id/empty"。第 2 行的 android:layout_width 属性用来设置 TextView 的宽度,wrap_content 表示 TextView 的宽度只要能够包含所显示的字符串即可。第 3 行的 android:layout_height 属性用来设置 TextView 的高度。第 4 行表示 TextView 所显示的字符串,在后面将通过代码更改 TextView 的显示内容。第 7 行中"fill_content"表示 EditText 的宽度将等于父控件的宽度。

TextViewDemo.java 文件中代码的修改如下。

```
1. TextView textView = (TextView)findViewById(R.id.TextView01);
2. EditText editText = (EditText)findViewById(R.id.EditText01);
3. textView.setText("用户名:");
4. editText.setText("");
```

如以上代码所示,第 1 行代码的 findViewById()函数能够通过 ID 引用界面上的任何控件,只要该控件在 XML 文件中定义过 ID 即可。第 3 行代码的 setText()函数用来设置 TextView 所显示的内容。

5.2.2 Button 和 ImageButton

Button 是一种按钮控件,用户能够在该控件上点击,并后引发相应的事件处理函数。ImageButton 用以实现能够显示图像功能的控件按钮,建立一个"ButtonDemo"的程序。如图 5-4 所示,包含 Button 和 ImageButton 两个按钮,上方是"Button 按钮",下方是一个 ImageButton 控件。

ButtonDemo 在 XML 文件中的代码如下。

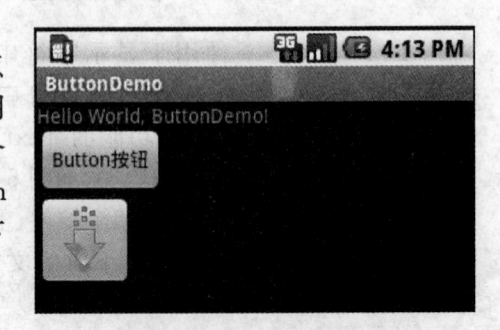

图 5-4 "ButtonDemo"视图效果

```
1. <Button android:id="@+id/Button01"
2.     android:layout_width="wrap_content"
3.     android:layout_height="wrap_content"
4.     android:text="Button01" >
5. </Button>
6. <ImageButton android:id="@+id/ImageButton01"
7.     android:layout_width="wrap_content"
8.     android:layout_height="wrap_content">
9. </ImageButton>
```

如以上代码所示,定义 Button 控件的高度、宽度和内容,定义 ImageButton 控件的高度和宽度,但是没定义显示的图像。在后面的代码中引入资源,将 download.png 文件拷贝到/res/drawable 文件夹下。如图 5-5 所示,在/res 目录上选择 Refresh,新添加的文件将显示在/res/drawable 文件夹下,R.java 文件内容也得到了更新,否则提示无法找到资源的错误。

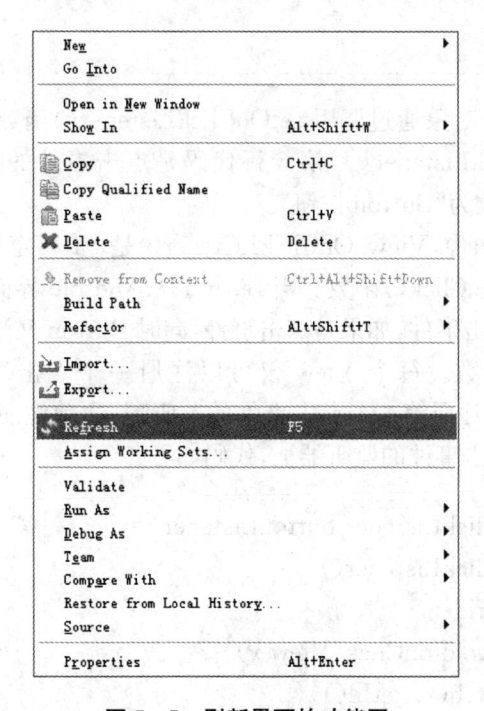

图 5-5 刷新界面的功能图

更改 Button 和 ImageButton 内容,引入 android.widget.Button 和 android.widget.ImageButton。

```
1. Button button = (Button)findViewById(R.id.Button01);
2. ImageButton imageButton = (ImageButton)findViewById(R.id.ImageButton01);
3. button.setText("Button 按钮");
4. imageButton.setImageResource(R.drawable.download);
```

如以上代码所示,第 1 行代码用于引用在 XML 文件中定义的 Button 控件,第 2 行代码用于引用在 XML 文件中定义的 ImageButton 控件,第 3 行代码将 Button 的显示内容更改为"Button 按钮",第 4 行代码利用 setImageResource()函数,将新加入的 png 文件 R. drawable. download 传递给 ImageButton。

为了使按钮响应点击事件,添加点击事件的监听器的代码如下。

```
1. final TextView textView = (TextView)findViewById(R. id. TextView01);
2. button. setOnClickListener(new View. OnClickListener() {
3.     public void onClick(View view) {
4.         textView. setText("Button 按钮");
5.     }
6. });
7. imageButton. setOnClickListener(new View. OnClickListener() {
8.     public void onClick(View view) {
9.         textView. setText("ImageButton 按钮");
10.    }
11. });
```

第 2 行代码中 button 对象通过调用 setOnClickListener()函数,注册一个点击(Click)事件的监听器 View. OnClickListener(),第 3 行代码是点击事件的回调函数,第 4 行代码将 TextView 的显示内容更改为"Button 按钮"。

View. OnClickListener(),View. OnClickListener()是 View 定义的点击事件的监听器接口,并在接口中仅定义了 onClick()函数。当 Button 从 Android 界面框架中接收到事件后,首先检查这个事件是否是点击事件,如果是点击事件,同时 Button 又注册了监听器,则会调用该监听器中的 onClick()函数。每个 View 仅可以注册一个点击事件的监听器,如果使用 setOnClickListener()函数注册第二个点击事件的监听器,之前注册的监听器将被自动注销。多个按钮注册到同一个点击事件的监听器上,代码如下。

```
1. Button. OnClickListener buttonListener = new
   Button. OnClickListener(){
2.     @Override
3.     public void onClick(View v) {
4.         switch(v. getId()){
5.         case R. id. Button01:
6.             textView. setText("Button 按钮");
7.             return;
8.         case R. id. ImageButton01:
9.             textView. setText("ImageButton 按钮");
10.            return;
11.        }
12. }};
```

```
13.    button.setOnClickListener(buttonListener);
14.    imageButton.setOnClickListener(buttonListener);
```

第 1 行至第 12 行代码定义了一个名为 buttonListener 的点击事件监听器,第 13 行代码将该监听器注册到 Button 上,第 14 行代码将该监听器注册到 ImageButton 上。

5.2.3 CheckBox 和 RadioButton

CheckBox 是一个同时可以选择多个选项的控件,RadioButton 则是仅可以选择一个选项的控件。RadioGroup 是 RadioButton 的承载体,程序运行时不可见,应用程序中可能包含一个或多个 RadioGroup,一个 RadioGroup 包含多个 RadioButton。在每个 RadioGroup 中,用户仅能够选择其中一个 RadioButton,建立一个"CheckboxRadiobuttonDemo"程序,如图 5-6 所示,包含五个控件,从上至下分别是 TextView01、CheckBox01、CheckBox02、RadioButton01、RadioButton02。当选择 RadioButton01,RadioButton02 则无法选择。

图 5-6 "CheckboxRadiobuttonDemo"视图效果

CheckboxRadiobuttonDemo 在 XML 文件中的代码如下。

```
1.  <TextView android:id="@+id/TextView01"
2.      android:layout_width="fill_parent"
3.      android:layout_height="wrap_content"
4.      android:text="@string/hello"/>
5.  <CheckBox android:id="@+id/CheckBox01"
6.      android:layout_width="wrap_content"
7.      android:layout_height="wrap_content"
8.      android:text="CheckBox01" >
9.  </CheckBox>
10. <CheckBox android:id="@+id/CheckBox02"
11.     android:layout_width="wrap_content"
12.     android:layout_height="wrap_content"
13.     android:text="CheckBox02" >
```

```
14.    </CheckBox>
15.        <RadioGroup android:id="@+id/RadioGroup01"
16.        android:layout_width="wrap_content"
17.        android:layout_height="wrap_content">
18.        <RadioButton android:id="@+id/RadioButton01"
19.            android:layout_width="wrap_content"
20.            android:layout_height="wrap_content"
21.            android:text="RadioButton01" >
22.        </RadioButton>
23.        <RadioButton android:id="@+id/RadioButton02"
24.            android:layout_width="wrap_content"
25.            android:layout_height="wrap_content"
26.            android:text="RadioButton02" >
27.        </RadioButton>
28. </RadioGroup>
```

第 15 行<RadioGroup>标签声明了一个 RadioGroup。在第 18 行和第 23 行分别声明了两个 RadioButton，这两个 RadioButton 是 RadioGroup 的子元素。引用 CheckBox 和 RadioButton 的方法参考下面的代码。

```
1. CheckBox checkBox1= (CheckBox)findViewById(R.id.CheckBox01);
2. RadioButton radioButton1
    =(RadioButton)findViewById(R.id.RadioButton01);
```

CheckBox 设置点击事件监听器的简要代码如下。

```
1. CheckBox.OnClickListener checkboxListener = new
   CheckBox.OnClickListener(){
2.   @Override
3.   public void onClick(View v) {
4.     //过程代码
5. }};
6. checkBox1.setOnClickListener(checkboxListener);
7. checkBox2.setOnClickListener(checkboxListener);
```

与 Button 设置点击事件监听器中介绍的方法相似，唯一不同在于将 Button.OnClickListener 换成了 CheckBox.OnClickListener。

RadioButton 设置点击事件监听器的方法，代码如下。

```
1. RadioButton.OnClickListener radioButtonListener = new
   RadioButton.OnClickListener(){
```

```
2.  @Override
3.  public void onClick(View v) {
4.     //过程代码
5.  }};
6.  radioButton1.setOnClickListener(radioButtonListener);
7.  radioButton2.setOnClickListener(radioButtonListener);
```

5.2.4 Spinner

Spinner 是一种能够从多个选项中选择其中一个的控件,类似于桌面程序的组合框(ComboBox),但没有组合框的下拉菜单,而是使用浮动菜单为用户提供选择。如图 5-7 所示,建立一个程序"SpinnerDemo",包含 3 个子项 Spinner 控件。

图 5-7 "SpinnerDemo"视图效果

SpinnerDemo 在 XML 文件中的代码如下。

```
1.  <TextView  android:id="@+id/TextView01"
2.      android:layout_width="fill_parent"
3.      android:layout_height="wrap_content"
4.      android:text="@string/hello"/>
5.  <Spinner android:id="@+id/Spinner01"
6.     android:layout_width="300dip"
7.     android:layout_height="wrap_content">
8.  </Spinner>
```

第 5 行使用<Spinner>标签声明了一个 Spinner 控件。第 6 行代码中指定了该控件的宽度为 300dip。在 SpinnerDemo.java 文件中,定义一个 ArrayAdapter 适配器,在 ArrayAdapter

中添加需要在 Spinner 中可以选择的内容,需要在代码中引入 android.widget.ArrayAdapter 和 android.widget.Spinner。

1. Spinner spinner = (Spinner) findViewById(R.id.Spinner01);
2. List<String> list = new ArrayList<String>();
3. list.add("Spinner 子项 1");
4. list.add("Spinner 子项 2");
5. list.add("Spinner 子项 3");
6. ArrayAdapter<String> adapter = new ArrayAdapter<String>(this, android.R.layout.simple_spinner_item, list);
7. adapter.setDropDownViewResource(android.R.layout.simple_spinner_dropdown_item);
8. spinner.setAdapter(adapter);

第 2 行代码建立了一个字符串数组列表(ArrayList),这种数组列表可以根据需要进行增减,<String>表示数组列表中保存的是字符串类型的数据。在代码的第 3、4、5 行中,使用 add()函数分别向数组列表中添加 3 个字符串。第 6 行代码建立了一个 ArrayAdapter 的数组适配器,数组适配器能够将界面控件和底层数据绑定在一起。第 7 行代码设定了 Spinner 的浮动菜单的显示方式,其中,android.R.layout.simple_spinner_dropdown_item 是 Android 系统内置的一种浮动菜单。第 8 行代码实现绑定过程,所有 ArrayList 中的数据,将显示在 Spinner 的浮动菜单中,设置 android.R.layout.simple_spinner_item 浮动菜单,显示结果如图 5-8 所示。

图 5-8 "SpinnerDemo"视图效果

适配器绑定界面控件和底层数据,如果底层数据更改了,用户界面也相应修改显示内容,就不需要应用程序再监视,从而极大地简化了代码的复杂性。

5.2.5 ListView

ListView 是一种用于垂直显示的列表控件,如果显示内容过多,则会出现垂直滚动条。

ListView 能够通过适配器将数据和自身绑定,在有限的屏幕上提供大量内容供用户选择,所以是经常使用的用户界面控件。ListView 支持点击事件处理,用户可以用少量的代码实现复杂的选择功能。下面建立一个"ListViewDemo"程序,包含四个控件,从上至下分别为 TextView01、ListView01、ListView02 和 ListView03。

ListViewDemo 在 XML 文件中的代码如下所示:

```
1.  <TextView  android:id="@+id/TextView01"
2.      android:layout_width="fill_parent"
3.      android:layout_height="wrap_content"
4.      android:text="@string/hello" />
5.  <ListView android:id="@+id/ListView01"
6.      android:layout_width="wrap_content"
7.      android:layout_height="wrap_content">
8.  </ListView>
```

在 ListViewDemo.java 文件中,首先需要为 ListView 创建适配器,并添加 ListView 中所显示的内容,代码如下所示:

```
1.  final TextView textView = 
       (TextView)findViewById(R.id.TextView01);
2.  ListView listView = (ListView)findViewById(R.id.ListView01);
3.  List<String> list = new ArrayList<String>();
4.  list.add("ListView 子项 1");
5.  list.add("ListView 子项 2");
6.  list.add("ListView 子项 3");
7.  ArrayAdapter<String> adapter = new ArrayAdapter<String>(this,
       android.R.layout.simple_list_item_1, list);
8.  listView.setAdapter(adapter);
```

第 2 行代码通过 ID 引用了 XML 文件中声明的 ListView,第 7 行代码声明了适配器 ArrayAdapter,第三个参数 list 说明适配器的数据源为数组列表,第 8 行代码将 ListView 和适配器绑定。

下面的代码声明了 ListView 子项的点击事件监听器,用以确定用户在 ListView 中,选择的是哪一个子项。

```
1.  AdapterView.OnItemClickListener listViewListener = new
       AdapterView.OnItemClickListener(){
2.      @Override
3.      public void onItemClick(AdapterView<?> arg0, View arg1, int arg2,
           long arg3) {
```

```
4.      String msg ="";
5.      textView.setText(msg);
6.  }};
7.  listView.setOnItemClickListener(listViewListener);
```

第 1 行的 AdapterView.OnItemClickListener 是 ListView 子项的点击事件监听器,同样是一个接口,需要实现 onItemClick()函数。在 ListView 子项被选择后,onItemClick()函数将被调用。第 3 行的 onItemClick()函数中一共有四个参数,参数 0 表示适配器控件,就是 ListView;参数 1 表示适配器内部的控件,是 ListView 中的子项;参数 2 表示适配器内部的控件,也就是子项的位置;参数 3 表示子项的行号。第 4 行和第 5 行代码用于显示信息,选择子项确定后,在 TextView 中显示子项父控件的信息、子控件信息、位置信息和 ID 信息。第 7 行代码是 ListView 指定刚刚声明的监听器。

5.2.6 TabHost

Tab 标签页是界面设计时经常使用的界面控件,可以实现多个分页之间的快速切换,每个分页可以显示不同的内容。如图 5-9 所示是 Android 系统内置的 Tab 标签页,点击"呼出/接听键"后出现,用于电话呼出和查看拨号记录、联系人。

图 5-9　Tab 标签页　　　　　　图 5-10　"TabDemo"视图效果

Tab 标签页的使用,首先要设计所有的分页的界面布局,在分页设计完成后,使用代码建立 Tab 标签页,并给每个分页添加标识和标题,最后确定每个分页所显示的界面布局。每个分页建立一个 XML 文件,用以编辑和保存分页的界面布局,使用的方法与设计普通用户界面没有什么区别。建立一个"TabDemo"程序,如图 5-10 所示,包含三个 XML 文件,分别为 tab1.xml、tab2.xml 和 tab3.xml。这 3 个文件分别使用线性布局、相对布局和绝对布局示例中 main.xml 的代码,并将布局的 ID 分别定义为 layout01、layout02 和 layout03。

tab1.xml 文件代码如下所示。

```
1. <?xml version="1.0" encoding="utf-8"?>
2. <LinearLayout android:id = "@+id/layout01"
```

```
3.    ……
4.    ……
5.  </LinearLayout>
```

tab2.xml 文件代码如下所示。

```
1.  <? xml version="1.0" encoding="utf-8"? >
2.  <AbsoluteLayout android:id="@+id/layout02"
3.    ……
4.    ……
5.  </AbsoluteLayout>
```

tab3.xml 文件代码如下所示。

```
1.  <? xml version="1.0" encoding="utf-8"? >
2.  <RelativeLayout android:id="@+id/layout03"
3.    ……
4.    ……
5.  </RelativeLayout>
```

在 TabDemo.java 文件中键入下面的代码,创建 Tab 标签页,并建立子页与界面布局直接的关联关系,代码如下所示。

```
1.  package edu.xzceu.TabDemo;
2.  import android.app.TabActivity;
3.  import android.os.Bundle;
4.  import android.widget.TabHost;
5.  import android.view.LayoutInflater;
6.
7.  public class TabDemo extends TabActivity {
8.      @Override
9.      public void onCreate(Bundle savedInstanceState) {
10.         super.onCreate(savedInstanceState);
11.         TabHost tabHost = getTabHost();
12.         LayoutInflater.from(this).inflate(R.layout.tab1,
                tabHost.getTabContentView(),true);
13.         LayoutInflater.from(this).inflate(R.layout.tab2,
                tabHost.getTabContentView(),true);
14.         LayoutInflater.from(this).inflate(R.layout.tab3,
                tabHost.getTabContentView(),true);
```

```
15.     tabHost.addTab(tabHost.newTabSpec("TAB1")
16.         .setIndicator("线性布局").setContent(R.id.layout01));
17.     tabHost.addTab(tabHost.newTabSpec("TAB2")
18.         .setIndicator("绝对布局").setContent(R.id.layout02));
19.     tabHost.addTab(tabHost.newTabSpec("TAB3")
20.         .setIndicator("相对布局").setContent(R.id.layout03));
21.   }
22. }
```

第 8 行代码的声明 TabDemo 类继承与 TabActivity，与以往继承 Activity 不同，TabActivity 支持内嵌多个 Activity 或 View。第 12 行代码通过 getTabHost() 函数获得了 Tab 标签页的容器，用以承载可以点击的 Tab 标签和分页的界面布局。第 13 行代码通过 LayoutInflater 将 tab1.xml 文件中的布局转换为 Tab 标签页可以使用的 View 对象。第 16 行代码使用 addTab() 函数添加了第 1 个分页，tabHost.newTabSpec("TAB1") 表明在第 12 行代码中建立的 tabHost 上，添加一个标识为 TAB1 的 Tab 分页。第 17 行代码使用 setIndicator() 函数设定分页显示的标题，使用 setContent() 函数设定分页所关联的界面布局。

TabDemo 示例的运行结果如图 5-11 所示。

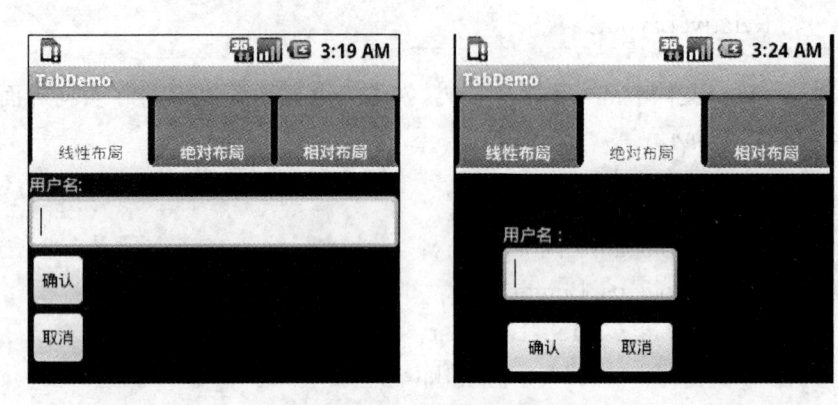

图 5-11 "TabDemo" 视图效果

在使用 Tab 标签页时，可以将不同分页的界面布局保存在不同的 XML 文件中，也可以将所有分页的布局保存在同一个 XML 文件中。第一种方法有利于在 Eclipse 开发环境中进行可视化设计，并且不同分页的界面布局在不同的文件中更加易于管理；第二种方法则可以产生较少的 XML 文件，同时编码时的代码也会更加简洁。

5.3 界面布局

界面布局(Layout)是用户界面结构的描述，定义了界面中所有的元素、结构和相互关系。声明 Android 程序的界面布局有两种方法：使用 XML 文件描述界面布局，在程序运行时动态添加或修改界面布局；使用 XML 文件声明界面布局的特点，将程序的表现层和控制层分离。用户既可以独立使用任何一种声明界面布局的方式，也可以同时使用两种方式。

在后期修改用户界面时,无需更改程序的源代码,用户还能够通过可视化工具直接看到所设计的用户界面,有利于加快界面设计的过程,并且为界面设计与开发带来极大的便利性。

5.3.1 线性布局

线性布局(LinearLayout)是一种重要的界面布局,也是经常使用到的一种界面布局,在线性布局中,所有的子元素都按照垂直或水平的顺序在界面上排列。如图 5-12 所示,如果垂直排列,则每行仅包含一个界面元素。如图 5-13 所示,如果水平排列,则每列仅包含一个界面元素。

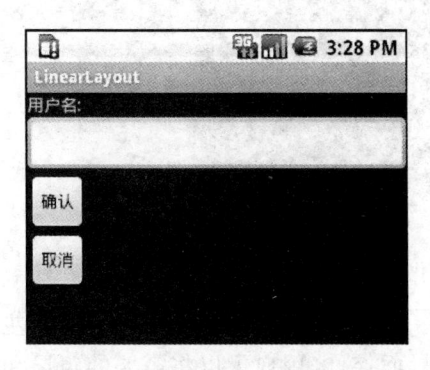

图 5-12 垂直排列效果图　　　　图 5-13 水平排列效果图

创建 Android 工程,工程名称是 LinearLayout,包名称是 edu.xzceu.LinearLayout,Activity 名称为 LinearLayout。为了能够完整体验创建线性布局的过程,首先删除 Eclipse 自动建立的/res/layout/main.xml 文件,然后建立用于显示垂直排列线性布局的 XML 文件。右击/res/layout 文件夹,如图 5-14 所示,选择 New → File 打开新文件建立向导,文件名为 main_vertical.xml,保存位置为 LinearLayout/res/layout。

图 5-14 "main_vertical.xml"的保存路径

如图 5-15 所示，双击新建立的/res/layout/main_vertical.xml 文件，Eclipse 将打开界面布局的可视化编辑器。

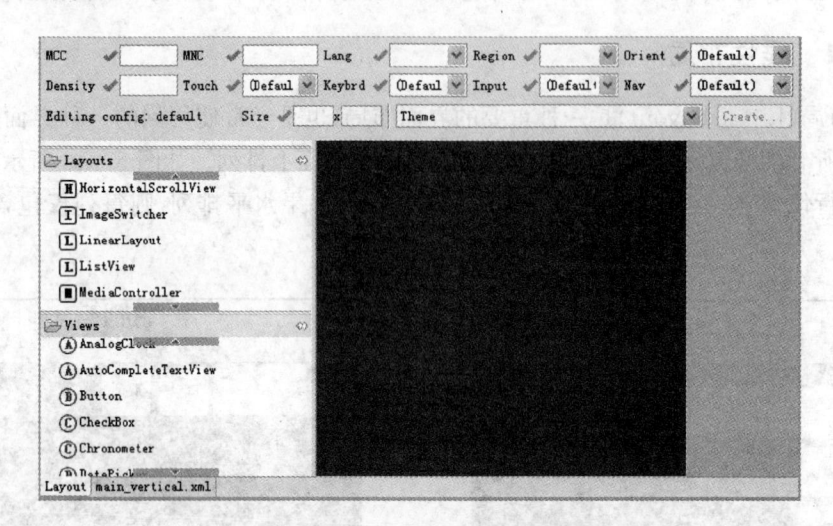

图 5-15　可视化编辑器

可视化编辑器顶部是资源配置清单，可以根据手机的配置不同选择不同的资源，主要用来实现应用软件的本地化。下部左侧是界面布局和界面控件，用户可以将需要的布局和控件拖拽到右面的可视化界面中，并修改布局和控件的属性。右侧是可视化的用户界面，能够实时地呈现用户界面，但无法正确显示中文。左下角的"Layout"和"main_vertical.xml"能够在可视化编辑器和 XML 文件编辑器之间切换。如图 5-16 所示，在 Eclipse 右边的 Outline 中，双击 LinearLayout，打开线性布局的属性编辑器，线性布局的排列方法主要由 Orientation 属性进行控制，vertical 表示垂直排列，horizontal 表示水平排列。选择 Orientation 的值为 vertical，表示该线性布局为垂直排列。

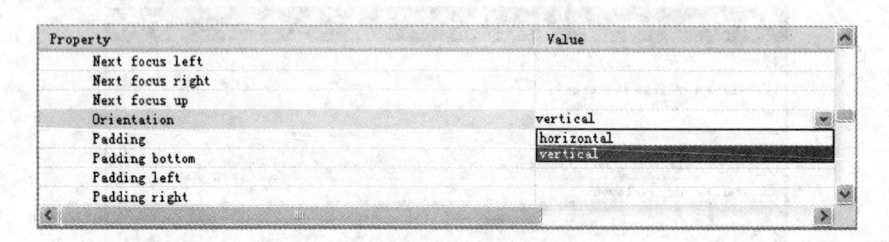

图 5-16　Orientation 属性选择

缺省情况下，Layout height 的值为 wrap_content，表示线性布局高度等于所有子控件的高度总和，也就是线性布局的高度会刚好将所有子控件包含其中。将 Layout width 属性的值改为 fill_parent，表示线性布局宽度等于父控件的宽度，就是将线性布局在横向上占据父控件的所有空间。

打开 XML 文件编辑器，main_vertical.xml 文件的代码如下。

```
1. <? xml version="1.0" encoding="utf-8"? >
2. <LinearLayout
3.    xmlns:android="http://schemas.android.com/apk/res/android"
```

```
4.    android:layout_width="fill_parent"
5.    android:layout_height="wrap_content"
6.    android:orientation="vertical">
7.  </LinearLayout>
```

第 2 行代码是声明 XML 文件的根元素为线性布局,第 4、5、6 行代码是在属性编辑器中修改过的宽度、高度和排列方式的属性。用户在可视化编辑器和属性编辑器中的任何修改,都会同步的反映在 XML 文件中,反之也是如此。

如图 5-17 所示,将四个界面控件 TextView、EditText、Button、Button 先后拖拽到可视化编辑器中,所有控件都自动获取控件名称,并把该名称显示在控件上,如 TextView01、EditText01、Button01 和 Button02。

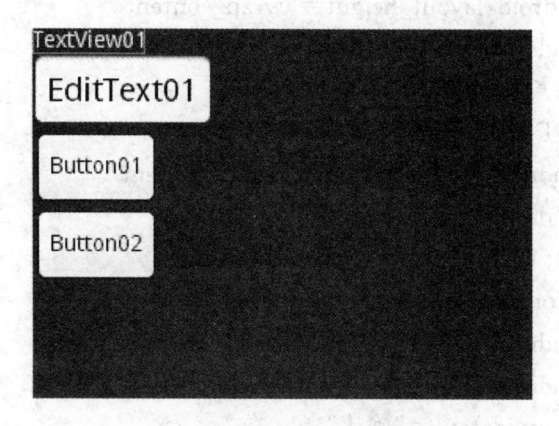

图 5-17 拖拽界面后效果图

修改界面控件的属性如表 5-1 所示。

表 5-1 界面控件的属性

编号	类型	属性	值
1	TextView	Id	@+id/label
		Text	用户名:
2	EditText	Id	@+id/entry
		Layout width	fill_parent
		Text	[null]
3	Button	Id	@+id/ok
		Text	确认
4	Button	Id	@+id/cancel
		Text	取消

所有界面控件都有一个共同的属性 ID,ID 是一个字符串,编译时被转换为整数,可以用来在代码中引用界面元素,一般仅在代码中需要动态修改的界面元素时才为界面元素设置 ID,反之则不需要设置 ID。从可视化编辑器中发现,界面控件的中文字符都显示为"□",因为可视化编辑

器还不能很好的支持中文字符。打开 XML 文件编辑器查看 main_vertical.xml 文件代码，发现在属性编辑器内填入的文字已经正常写入到 XML 文件中，例如第 11、20、25 行的代码。

```
1.  <?xml version="1.0" encoding="utf-8"?>
2.  <LinearLayout
3.      xmlns:android="http://schemas.android.com/apk/res/android"
4.      android:layout_width="fill_parent"
5.      android:layout_height="wrap_content"
6.      android:orientation="vertical">
7.  <TextView android:id="@+id/label"
8.      android:layout_width="wrap_content"
9.      android:layout_height="wrap_content"
10.     android:text="用户名:" >
11. </TextView>
12. <EditText android:id="@+id/entry"
13.     android:layout_height="wrap_content"
14.     android:layout_width="fill_parent">
15. </EditText>
16. <Button android:id="@+id/ok"
17.     android:layout_width="wrap_content"
18.     android:layout_height="wrap_content"
19.     android:text="确认">
20. </Button>
21. <Button android:id="@+id/cancel"
22.     android:layout_width="wrap_content"
23.     android:layout_height="wrap_content"
24.     android:text="取消" >
25. </Button>
26. </LinearLayout>
```

将 LinearLayout.java 文件中的 setContentView(R.layout.main)更改为 setContentView(R.layout.main_vertical)。运行后的结果如图 5-18 所示。

建立横向线性布局与纵向线性布局相似，只需注意以下几点：建立的 main_horizontal.xml 文件线性布局的 Orientation 属性的值设置为 horizontal，将 EditText 的 Layout width 属性的值设置为 wrap_content，将 LinearLayout.java 文件中的 setContentView(R.layout.main_vertical)修改为 setContentView(R.layout.main_horizontal)。

图 5-18　更改路径后效果图

5.3.2 框架布局

框架布局(FrameLayout)是最简单的界面布局,是用来存放一个元素的空白空间,且子元素的位置是不能够指定的,只能够放置在空白空间的左上角。如果有多个子元素,后放置的子元素将遮挡先放置的子元素。使用 Android SDK 中提供的层级观察器(Hierarchy Viewer)进一步分析界面布局。层级观察器能够对用户界面进行分析和调试,并以图形化的方式展示树形结构的界面布局,还提供了一个精确的像素级观察器(Pixel Perfect View),以栅格的方式详细观察放大后的界面。如图 5-19 所示,在层级观察器中获得示例界面布局的树形结构图。

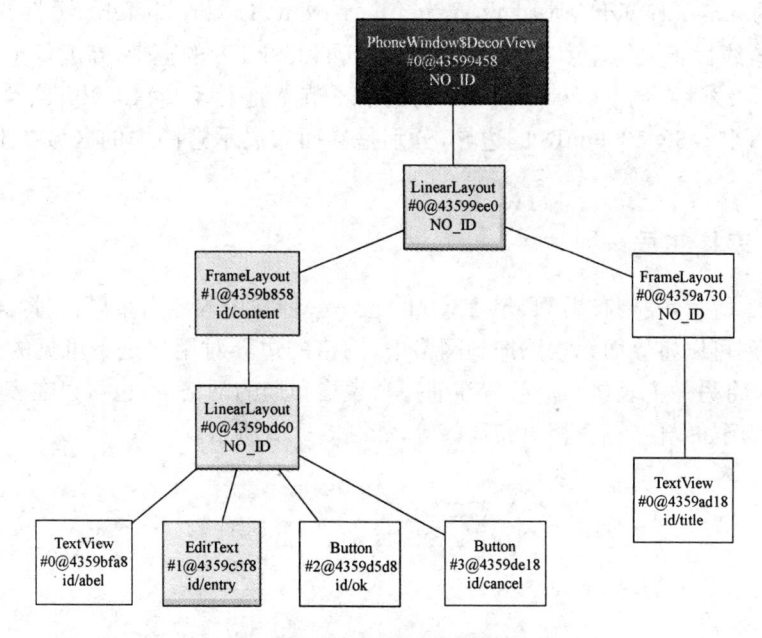

图 5-19 树形结构图

图 5-20 示意图

结合界面布局的树形结构图(图 5-19)和示意图(图 5-20),分析不同界面布局和界面控件的区域边界。用户界面的根节点(♯0@43599ee0)是线性布局,其边界是整个界面,也就是示意图的最外层的实心线。根节点右侧的子节点(♯0@43599a730)是框架布局,仅有一个节点元素(♯0@4359ad18),这个子元素是 TextView 控件,用来显示 Android 应用程序名称,其边界是示意图中的区域 1。因此框架布局元素♯0@43599a730 的边界是同区域 1 的高度相同,宽带充满整个根节点的区域。这两个界面元素是系统自动生成的,一般情况下用户不能够修改和编辑。根节点左侧的子节点(♯1@4359b858)也是框架布局,边界是区域 2 到区域 7 的全部空间。子节点(♯1@4359b858)下仅有一个子节点(♯0@4359bd60)元素是线性布局,因为线性布局的 Layout width 属性设置为 fill_parent,Layout height 属性设置为 wrap_content,因此该线性布局的宽度就是其父节点♯1@4359b858 的宽带,高度等于所有子节点元素的高度之和。线性布局♯0@4359bd60 的四个子节点元素♯0@4359bfa8、♯1@4359c5f8、♯2@4359d5d8 和♯3@4359de18 的边界,分别是界面布局示意图中的区域 2、区域 3、区域 4 和区域 5

5.3.3 表格布局

如图 5-21 所示,表格布局(TableLayout)也是一种常用的界面布局,它将屏幕划分网格,通过指定行和列可以将界面元素添加到网格中,网格的边界对用户是不可见的。表格布局还支持嵌套,可以将另一个表格布局放置在前一个表格布局的网格中,也可以在表格布局中添加其他界面布局,例如线性布局、相对布局等等,如图 5-22 所示。

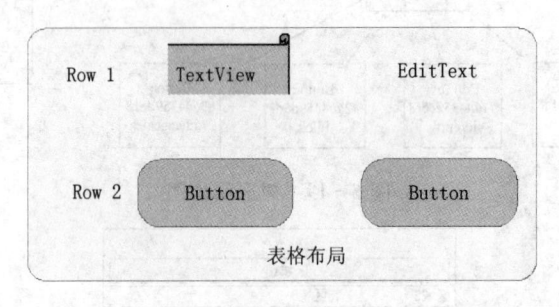

图 5-21 表格布局示意图

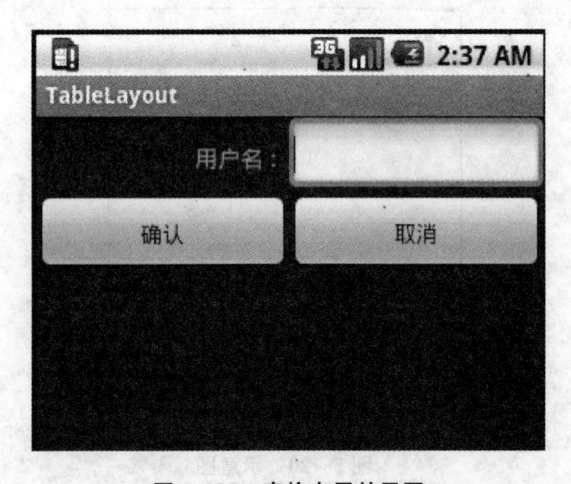

图 5-22 表格布局效果图

第 5 章 Android 用户界面

如图 5-23 所示,建立表格布局要注意以下几点,向界面中添加一个线性布局,无需修改布局的属性值。其中,Id 属性为 TableLayout01,Layout width 和 Layout height 属性都为 wrap_content,向 TableLayout01 中添加两个 TableRow。TableRow 代表一个单独的行,每行被划分为几个小的单元,单元中可以添加一个界面控件。其中,Id 属性分别为 TableRow01 和 TableRow02,Layout width 和 Layout height 属性都为 wrap_content。

通过 Outline,向 TableRow01 中添加 TextView 和 EditText,向 TableRow02 中添加两个 Button。

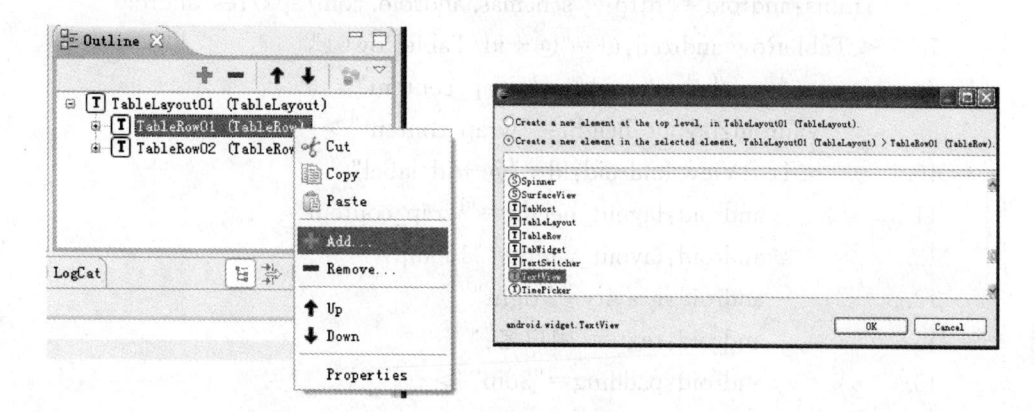

图 5-23 建立表格布局

参考下表设置 TableRow 中四个界面控件的属性值,如下表 5-2 所示。

表 5-2 控件的属性值

编 号	类 型	属 性	值
1	TextView	Id	@+id/label
		Text	用户名:
		Gravity	right
		Padding	3dip
		Layout width	160dip
2	EditText	Id	@+id/entry
		Text	[null]
		Padding	3dip
		Layout width	160dip
3	Button	Id	@+id/ok
		Text	确认
		Padding	3dip
4	Button	Id	@+id/cancel
		Text	取消
		Padding	3dip

main.xml 文件的完整代码如下。

```xml
1.  <?xml version="1.0" encoding="utf-8"?>
2.
3.  <TableLayout android:id="@+id/TableLayout01"
4.      android:layout_width="fill_parent"
5.      android:layout_height="fill_parent"
6.      xmlns:android="http://schemas.android.com/apk/res/android">
7.    <TableRow android:id="@+id/TableRow01"
8.        android:layout_width="wrap_content"
9.        android:layout_height="wrap_content">
10.      <TextView android:id="@+id/label"
11.          android:layout_height="wrap_content"
12.          android:layout_width="160dip"
13.          android:gravity="right"
14.          android:text="用户名:"
15.          android:padding="3dip">
16.      </TextView>
17.      <EditText android:id="@+id/entry"
18.          android:layout_height="wrap_content"
19.          android:layout_width="160dip"
20.          android:padding="3dip">
21.      </EditText>
22.    </TableRow>
23.    <TableRow android:id="@+id/TableRow02"
24.        android:layout_width="wrap_content"
25.        android:layout_height="wrap_content">
26.      <Button android:id="@+id/ok"
27.          android:layout_height="wrap_content"
28.          android:padding="3dip"
29.          android:text="确认">
30.      </Button>
31.      <Button android:id="@+id/Button02"
32.          android:layout_width="wrap_content"
33.          android:layout_height="wrap_content"
34.          android:padding="3dip"
35.          android:text="取消">
36.      </Button>
37.    </TableRow>
38.  </TableLayout>
```

如上代码所示，第3行代码使用了＜TableLayout＞标签声明表格布局。第7行和第23行代码声明了两个 TableRow 元素。第12行设定宽度属性 android:layout_width:160dip。第13行设定属性 android:gravity,指定文字为右对齐。第15行使用属性 android:padding,声明 TextView 元素与其他元素的间隔距离为 3dip。

相对布局(RelativeLayout)是一种非常灵活的布局方式,能够通过指定界面元素与其他元素的相对位置关系,确定界面中所有元素的布局位置,能够最大程度保证在各种屏幕类型的手机上正确显示界面布局。

在使用相对布局示例时,添加 TextView 控件("用户名"),相对布局会将 TextView 控件放置在屏幕的最上方；然后添加 EditText 控件(输入框),并声明该控件的位置在 TextView 控件的下方,相对布局会根据 TextView 的位置确定 EditText 控件的位置；之后添加第一个 Button 控件("取消"按钮),声明在 EditText 控件的下方,且在父控件的最右边；最后,添加第二个 Button 控件("确认"按钮),声明该控件在第一个 Button 控件的左方,且与第一个 Button 控件处于相同的水平位置。相对布局在 main.xml 文件的完整代码如下。

```
1.  <?xml version="1.0" encoding="utf-8"?>
2.
3.  <RelativeLayout android:id="@+id/RelativeLayout01"
4.      android:layout_width="fill_parent"
5.      android:layout_height="fill_parent"
6.      xmlns:android="http://schemas.android.com/apk/res/android">
7.      <TextView android:id="@+id/label"
8.          android:layout_height="wrap_content"
9.          android:layout_width="fill_parent"
10.         android:text="用户名:">
11.     </TextView>
12.     <EditText android:id="@+id/entry"
13.         android:layout_height="wrap_content"
14.         android:layout_width="fill_parent"
15.         android:layout_below="@id/label">
16.     </EditText>
17.     <Button android:id="@+id/cancel"
18.         android:layout_height="wrap_content"
19.         android:layout_width="wrap_content"
20.         android:layout_alignParentRight="true"
21.         android:layout_marginLeft="10dip"
22.         android:layout_below="@id/entry"
23.         android:text="取消">
24.     </Button>
25.     <Button android:id="@+id/ok"
```

```
26.    android:layout_height="wrap_content"
27.    android:layout_width="wrap_content"
28.    android:layout_toLeftOf="@id/cancel"
29.    android:layout_alignTop="@id/cancel"
30.    android:text="确认">
31. </Button>
32. </RelativeLayout>
```

如上所示：第 3 行使用了<RelativeLayout>标签声明一个相对布局，第 15 行使用位置属性 android:layout_below，确定 EditText 控件在 ID 为 label 的元素下方，第 20 行使用属性 android:layout_alignParentRight，声明该元素在其父元素的右边边界对齐，第 21 行设定属性 android:layout_marginLeft，左移 10dip，第 22 行声明该元素在 ID 为 entry 的元素下方，第 28 行声明使用属性 android:layout_toLeftOf，声明该元素在 ID 为 cancel 元素的左边，第 29 行使用属性 android:layout_alignTop，声明该元素与 ID 为 cancel 的元素在相同的水平位置。

绝对布局（AbsoluteLayout）能通过指定界面元素的坐标位置，来确定用户界面的整体布局。绝对布局是一种不推荐使用的界面布局，因为通过 X 轴和 Y 轴确定界面元素位置后，Android 系统不能够根据不同屏幕对界面元素的位置进行调整，降低了界面布局对不同类型和尺寸屏幕的适应能力，每一个界面控件都必须指定坐标(X,Y)。如图 5-24 所示，"确认"按钮的坐标是(40,120)，"取消"按钮的坐标是(120,120)，坐标原点(0,0)在屏幕的左上角。

图 5-24 绝对布局视图效果

绝对布局示例在 main.xml 文件的完整代码如下。

```
1. <?xml version="1.0" encoding="utf-8"?>
2.
3. <AbsoluteLayout android:id="@+id/AbsoluteLayout01"
4.    android:layout_width="fill_parent"
5.    android:layout_height="fill_parent"
6.    xmlns:android="http://schemas.android.com/apk/res/android">
7. <TextView android:id="@+id/label"
8.    android:layout_x="40dip"
```

```
9.        android:layout_y="40dip"
10.       android:layout_height="wrap_content"
11.       android:layout_width="wrap_content"
12.       android:text="用户名:">
13. </TextView>
14. <EditText android:id="@+id/entry"
15.       android:layout_x="40dip"
16.       android:layout_y="60dip"
17.       android:layout_height="wrap_content"
18.       android:layout_width="150dip">
19. </EditText>
20. <Button android:id="@+id/ok"
21.       android:layout_width="70dip"
22.       android:layout_height="wrap_content"
23.       android:layout_x="40dip"
24.       android:layout_y="120dip"
25.       android:text="确认">
26. </Button>
27. <Button android:id="@+id/cancel"
28.       android:layout_width="70dip"
29.       android:layout_height="wrap_content"
30.       android:layout_x="120dip"
31.       android:layout_y="120dip"
32.       android:text="取消">
33. </Button>
34. </AbsoluteLayout>
```

5.4 菜单

菜单是应用程序中非常重要的组成部分,能够在不占用界面空间的前提下,为应用程序提供统一的功能和设置界面,并为程序开发人员提供了易于使用的编程接口。Android 系统支持三种菜单:选项菜单(Option Menu),子菜单(Submenu)和快捷菜单(Context Menu)。

5.4.1 选项菜单

选项菜单是一种经常被使用的 Android 系统菜单,如图 5-25 所示,通过"菜单键"(MENU key)打开。选项菜单分为图标菜单(Icon Menu)和扩展菜单(Expanded Menu),图标菜单能够同时显示文字和图标的菜单,最多支持 6 个子项,图标菜单不支持单选框和复选框。

图 5-25 "选项菜单"视图效果

扩展菜单是在图标菜单子项多余 6 个时才出现,通过点击图标菜单最后的子项"More"才能打开,如图 5-26 所示。扩展菜单是垂直的列表型菜单,不能够显示图标,支持单选框和复选框。

图 5-26 "扩展菜单"视图效果

重载 Activity 的 onCreateOptionMenu()函数,才能够在 Android 应用程序中使用选项菜单。初次使用选项菜单时,会调用 onCreateOptionMenu()函数,用来初始化菜单子项的相关内容,设置菜单子项自身的子项的 ID 和组 ID。菜单子项显示的文字和图片等。

```
1.   final static int MENU_DOWNLOAD = Menu.FIRST;
2.   final static int MENU_UPLOAD = Menu.FIRST+1;
3.   @Override
4.   public boolean onCreateOptionsMenu(Menu menu){
5.       menu.add(0,MENU_DOWNLOAD,0,"下载设置");
6.       menu.add(0,MENU_UPLOAD,1,"上传设置");
7.       return true;
8.   }
```

如上所示,第 1 行和第 2 行代码将菜单子项 ID 定义成静态常量,并使用静态常量 Menu.FIRST(整数类型,值为 1)定义第一个菜单子项,以后的菜单子项仅需在 Menu.FIRST 增加相应的数值即可。第 4 行代码 Menu 对象作为一个参数被传递到函数内部,因此在 onCreateOptionsMenu()函数中,用户可以使用 Menu 对象的 add()函数添加菜单子项。第 7 行代码是 onCreateOptionsMenu()函数返回值,函数的返回值类型为布尔型,返回 true 将显示在函数中设置的菜单,否则不能够显示菜单。

add()函数的语法如下。

```
MenuItem android.view.Menu.add(int groupId, int itemId, int order, CharSequence title)
```

第 1 个参数 groupId 是组 ID,用以批量的对菜单子项进行处理和排序。第 2 个参数 itemId 是子项 ID,是每一个菜单子项的唯一标识,通过子项 ID 使应用程序能够定位到用户所选择的菜单子项。第 3 个参数 order 是定义菜单子项在选项菜单中的排列顺序。第 4 个参数 title 是菜单子项所显示的标题,添加菜单子项的图标和快捷键:使用 setIcon()函数和 setShortcut()函数,代码如下。

```
1.    menu.add(0,MENU_DOWNLOAD,0,"下载设置")
2.    .setIcon(R.drawable.download);
3.    .setShortcut(',','d');
```

MENU_DOWNLOAD 菜单设置图标和快捷键的代码。第 2 行代码中使用了新的图像资源,用户将需要使用的图像文件拷贝到/res/drawable 目录下。setShortcut()函数第一个参数是为数字键盘设定的快捷键,第二个参数是为全键盘设定的快捷键,且不区分字母的大小写。

重载 onPrepareOptionsMenu()函数,能够动态的添加、删除菜单子项,或修改菜单的标题、图标和可见性等内容。onPrepareOptionsMenu()函数的返回值的含义与 onCreateOptionsMenu()函数相同,返回 true 则显示菜单,返回 false 则不显示菜单。

下面的代码是在用户每次打开选项菜单时,在菜单子项中显示用户打开该子项的次数。

```
1.    static int MenuUploadCounter = 0;
2.    @Override
3.    public boolean onPrepareOptionsMenu(Menu menu){
4.        MenuItem uploadItem = menu.findItem(MENU_UPLOAD);
5.        uploadItem.setTitle("上传设置:" +String.valueOf(MenuUploadCounter));
6.        return true;
7.    }
```

第 1 行代码设置一个菜单子项的计数器,用来统计用户打开"上传设置"子项的次数。第 4 行代码是通过将菜单子项的 ID 传递给 menu.findItem()函数,获取到菜单子项的对象。第 5 行代码是通过 MenuItem 的 setTitle()函数修改菜单标题,onOptionsItemSelected()函数能够处理菜单选择事件,且该函数在每次点击菜单子项时都会被调用。下面的代码说明了如何通过菜单子项的子项 ID 执行不同的操作。

```
1.  @Override
2.  public boolean onOptionsItemSelected(MenuItem item){
3.      switch(item.getItemId()){
4.          case MENU_DOWNLOAD:
5.              MenuDownlaodCounter++;
6.              return true;
7.          case MENU_UPLOAD:
8.              MenuUploadCounter++;
9.              return true;
10.     }
11.     return false;
12. }
```

如上所示，onOptionsItemSelected()的返回值表示是否对菜单的选择事件进行处理，如果已经处理过则返回 true，否则返回 false。第 2 行的 MenuItem.getItemId()函数可以获取到被选择菜单子项的 ID，完整代码请参考 OptionsMenu 程序。如图 5-27 所示，程序运行后，通过点击"菜单键"可以调出程序设计的两个菜单子项。

图 5-27 "菜单键"视图效果　　　　图 5-28 "浮动窗口"视图效果

5.4.2 子菜单

子菜单是能够显示更加详细信息的菜单子项，如图 5-28 所示，菜单子项使用了浮动窗体的显示形式，能够更好适应小屏幕的显示方式。

Android 系统的子菜单使用非常灵活，可以在选项菜单或快捷菜单中使用子菜单，有利于将相同或相似的菜单子项组织在一起，便于显示和分类。子菜单不支持嵌套，子菜单的添加是使用 addSubMenu()函数实现。

1. SubMenu uploadMenu = (SubMenu) menu.addSubMenu(0, MENU_UPLOAD,1,"上传设置").setIcon(R.drawable.upload);
2. uploadMenu.setHeaderIcon(R.drawable.upload);
3. uploadMenu.setHeaderTitle("上传参数设置");
4. uploadMenu.add(0,SUB_MENU_UPLOAD_A,0,"上传参数 A");
5. uploadMenu.add(0,SUB_MENU_UPLOAD_B,0,"上传参数 B");

如上所示，第 1 行代码在 onCreateOptionsMenu()函数传递的 menu 对象上调用 addSubMenu()函数，在选项菜单中添加一个菜单子项，用户点击后可以打开子菜单，addSubMenu()函数与选项菜单中使用过的 add()函数支持相同的参数，同样可以指定菜单子项的 ID、组 ID 和标题等参数，并且能够通过 setIcon()函数菜单所显示的图标。第 2 行代码使用 setHeaderIcon()函数，定义子菜单的图标。第 3 行定义子菜单的标题，若不规定子菜单的标题，子菜单将显示父菜单子项标题，即第 1 行代码中"上传设置"。第 4 行和第 5 行在子菜单中添加了两个菜单子项，菜单子项的更新函数和选择事件处理函数，仍然使用 onPrepareOptionsMenu()函数和 onOptionsItemSelected()函数。

以上小节的代码为基础，如图 5-29 所示，将"上传设置"改为子菜单，并在子菜单中添加"上传参数 A"和"上传参数 B"两个菜单子项。完整代码请参考 MySubMenu 程序。

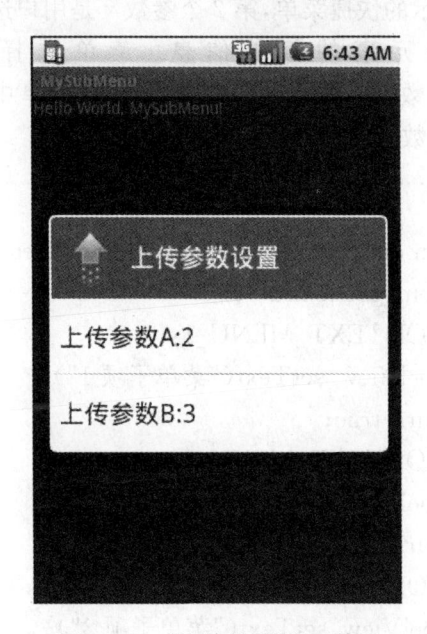

图 5-29 "上传设置"视图效果

5.4.3 快捷菜单

快捷菜单同样采用了动窗体的显示方式，与子菜单的实现方式相同，但两种菜单的启动方式却截然不同，启动方式：快捷菜单类似于普通桌面程序中的"右键菜单"，当用户点击界面元素超过 2 秒后，将启动注册到该界面元素的快捷菜单。使用方法：与使用选项菜单的方法非常相似，需要重载 onCreateContextMenu()函数和 onContextItemSelected()函数。onCreateContextMenu()函数主要用来添加快捷菜单所显示的标题、图标和菜单子项等内容，选项菜单中的 onCreateOptionsMenu()函数仅在选项菜单第一次启动时被调用一次,快捷菜单的 onCreateContextMenu()函数每次启动时都会被调用一次。

```
1.  final static int CONTEXT_MENU_1 = Menu.FIRST;
2.  final static int CONTEXT_MENU_2 = Menu.FIRST+1;
3.  final static int CONTEXT_MENU_3 = Menu.FIRST+2;
4.  @Override
5.  public void onCreateContextMenu(ContextMenu menu, View v,
        ContextMenuInfo menuInfo){
6.      menu.setHeaderTitle("快捷菜单标题");
7.      menu.add(0, CONTEXT_MENU_1, 0,"菜单子项1");
8.      menu.add(0, CONTEXT_MENU_2, 1,"菜单子项2");
9.      menu.add(0, CONTEXT_MENU_3, 2,"菜单子项3");
10. }
```

如上所示,ContextMenu 类支持 add()函数(代码第 7 行)和 addSubMenu()函数,可以在快捷菜单中添加菜单子项和子菜单。第 5 行代码的 onCreateContextMenu()函数中的参数,第 1 个参数 menu 是需要显示的快捷菜单,第 2 个参数 v 是用户选择的界面元素,第 3 个参数 menuInfo 是所选择界面元素的额外信息。菜单选择事件的处理需要重载 onContextItemSelected()函数,该函数在用户选择快捷菜单中的菜单子项后被调用,与 onOptionsItemSelected()函数的使用方法基本相同。

```
1.  @Override
2.  public boolean onContextItemSelected(MenuItem item){
3.      switch(item.getItemId()){
4.          case CONTEXT_MENU_1:
5.              LabelView.setText("菜单子项1");
6.              return true;
7.          case CONTEXT_MENU_2:
8.              LabelView.setText("菜单子项2");
9.              return true;
10.         case CONTEXT_MENU_3:
11.             LabelView.setText("菜单子项3");
12.             return true;
13.     }
14.     return false;
15. }
```

如上所示,使用 registerForContextMenu()函数,将快捷菜单注册到界面控件上(下方代码第 7 行),这样用户在长时间点击该界面控件时,便会启动快捷菜单。为了能够在界面上直接显示用户所选择快捷菜单的菜单子项,在代码中引用了界面元素 TextView(下方代码第 6 行),通过更改 TextView 的显示内容(上方代码第 5、8 和 11 行),显示用户所选择的菜单子项。

```
1.  TextView LabelView = null;
2.  @Override
3.  public void onCreate(Bundle savedInstanceState) {
4.      super.onCreate(savedInstanceState);
5.      setContentView(R.layout.main);
6.      LabelView = (TextView)findViewById(R.id.label);
7.      registerForContextMenu(LabelView);
8.  }
```

下方代码是/src/layout/main.xml 文件的部分内容,第 1 行声明了 TextView 的 ID 为 label,在上方代码的第 6 行中,通过 R.id.label 将 ID 传递给 findViewById()函数,这样用户便能够引用该界面元素,并能够修改该界面元素的显示内容。

```
1.  <TextView   android:id="@+id/label"
2.      android:layout_width="fill_parent"
3.      android:layout_height="fill_parent"
4.      android:text="@string/hello"
5.  />
```

需要注意的一点,上方代码的第 2 行,将 android:layout_width 设置为 fill_parent,这样 TextView 将填充满父节点的所有剩余屏幕空间,用户点击屏幕 TextView 下方任何位置都可以启动快捷菜单。如果将 android:layout_width 设置为 wrap_content,则用户必须准确点击 TextView 才能启动快捷菜单。完整代码参考 MyContextMenu 程序,如图 5-30 所示为运行结果。

图 5-30 "MyContextMenu"视图效果

在 Android 系统中,菜单不仅能够在代码中定义,而且可以像界面布局一样在 XML 文件中进行定义。使用 XML 文件定义界面菜单,将代码与界面设计分类,有助于简化代码的复杂程度,并且更有利于界面的可视化。下面将快捷菜单的示例程序 MyContextMen 改用 XML 实现,新程序的工程名称为 MyXMLContextMenu。

首先需要创建保存菜单内容的 XML 文件,在/src 目录下建立子目录 menu,并在 menu 下建立 context_menu.xml 文件,代码如下:

```
1. <menu xmlns:android="http://schemas.android.com/apk/res/android">
2. <item android:id="@+id/contextMenu1"
3.     android:title="菜单子项 1"/>
4. <item android:id="@+id/contextMenu2"
5.     android:title="菜单子项 2"/>
6. <item android:id="@+id/contextMenu3"
7.     android:title="菜单子项 3"/>
8. </menu>
```

在描述菜单的 XML 文件中,必须以<menu>标签(代码第 1 行)作为根节点,<item>标签(代码第 2 行)用来描述菜单中的子项。<item>标签可以通过嵌套实现子菜单的功能。

XML 菜单的显示结果如图 5-31 所示。

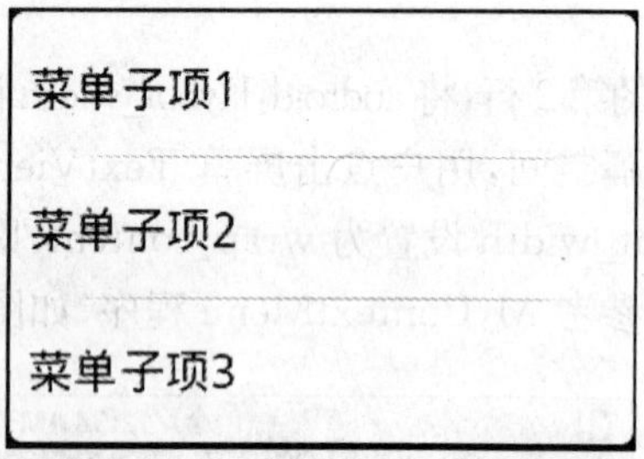

图 5-31　XML 菜单视图效果

在 XML 文件中定义菜单后,采用在 onCreateContextMenu()函数中调用 inflater.inflate()方法,将 XML 资源文件传递给菜单对象。

```
1. @Override
2. public void onCreateContextMenu(ContextMenu menu,
3.     View v, ContextMenuInfo menuInfo){
4.     MenuInflater inflater = getMenuInflater();
5.     inflater.inflate(R.menu.context_menu, menu);
6. }
```

如上代码,第 4 行代码中的 getMenuInflater()为当前的 Activity 返回 MenuInflater。第 5 行代码将 XML 资源文件 R.menu.context_menu,传递给 menu 这个快捷菜单对象。在 Android 系统中,存在多种界面事件,如点击事件、触摸事件、焦点事件和菜单事件等等。在这些界面事件发生时,Android 界面框架调用界面控件的事件处理函数对事件进行处理。

5.5 界面事件

5.5.1 按键事件

在 MVC 模型中,控制器根据界面事件(UI Event)的类型,将事件传递给界面控件不同的事件处理函数。按键事件(KeyEvent)将传递给 onKey( )函数进行处理,触摸事件(TouchEvent)将传递给 onTouch()函数进行处理。Android 系统界面事件的传递和处理遵循一定的规则,如果界面控件设置了事件监听器,则事件将先传递给事件监听器,如果界面控件没有设置事件监听器,界面事件则会直接传递给界面控件的其他事件处理函数,即使界面控件设置了事件监听器,界面事件也可以再次传递给其他事件处理函数。

Android 系统界面事件的传递和处理遵循一定的规则,是否继续传递事件给其他处理函数是由事件监听器处理函数的返回值决定的。如果监听器处理函数的返回值为 true,表示该事件已经完成处理过程,不需要其他处理函数参与处理过程,这样事件就不会再继续进行传递。如果监听器处理函数的返回值为 false,则表示该事件没有完成处理过程,或需要其他处理函数捕获到该事件,事件会被传递给其他的事件处理函数。以 EditText 控件中的按键事件为例,说明 Android 系统界面事件传递和处理过程,假设 EditText 控件已经设置了按键事件监听器。

当用户按下键盘上的某个按键时,控制器将产生 KeyEvent 按键事件,Android 系统会首先判断 EditText 控件是否设置了按键事件监听器,因为 EditText 控件已经设置按键事件监听器 OnKeyListener,所以按键事件先传递到监听器的事件处理函数 onKey()中。事件能够继续传递给 EditText 控件的其他事件处理函数,完全根据 onKey()函数的返回值来确定,如果 onKey()函数返回 false,事件将继续传递,这样 EditText 控件就可以捕获到该事件,将按键的内容显示在 EditText 控件中;如果 onKey()函数返回 true,将阻止按键事件的继续传递,这样 EditText 控件就不能够捕获到按键事件,也就不能够将按键内容显示在 EditText 控件中。Android 界面框架支持对按键事件的监听,并能够将按键事件的详细信息传递给处理函数。为了处理控件的按键事件,先需要设置按键事件的监听器,并重载 onKey()函数。

示例代码如下:

```
1. entryText.setOnKeyListener(new OnKeyListener(){
2.    @Override
3.    public boolean onKey(View view, int keyCode, KeyEvent keyEvent) {
4.        //过程代码……
5.        return true/false;
6.    }
```

第 1 行代码是设置控件的按键事件监听器。第 3 行代码的 onKey()函数中的参数,第 1 个参数 view 表示产生按键事件的界面控件,第 2 个参数 keyCode 表示按键代码,第 3 个参数 keyEvent 则包含了事件的详细信息,如按键的重复次数、硬件编码和按键标志等。第 5 行代码是 onKey()函数的返回值,返回 true,阻止事件传递,返回 false,允许继续传递按键事件。

KeyEventDemo 是一个说明如何处理按键事件的示例。KeyEventDemo 用户界面,最上方的 EditText 控件是输入字符的区域,中间的 CheckBox 控件用来控制 onKey()函数的返回值,最下方的 TextView 控件用来显示按键事件的详细信息,包括按键动作、按键代码、按键字符、Unicode 编码、重复次数、功能键状态、硬件编码和按键标志。

界面的 XML 文件的代码如下。

```
1. <EditText android:id="@+id/entry"
2.     android:layout_width="fill_parent"
3.     android:layout_height="wrap_content">
4. </EditText>
5. <CheckBox android:id="@+id/block"
6.     android:layout_width="wrap_content"
7.     android:layout_height="wrap_content"
8.     android:text="返回 true,阻止将按键事件传递给界面元素" >
9. </CheckBox>
10. <TextView android:id="@+id/label"
11.     android:layout_width="wrap_content"
12.     android:layout_height="wrap_content"
13.     android:text="按键事件信息" >
14. </TextView>
```

在 EditText 中,每当任何一个键被按下或抬起时,都会引发按键事件。为了能够使 EditText 处理按键事件,需要使用 setOnKeyListener()函数在代码中设置按键事件监听器,并在 onKey()函数添加按键事件的处理过程。

```
1. entryText.setOnKeyListener(new OnKeyListener(){
2.     @Override
3.     public boolean onKey(View view, int keyCode, KeyEvent keyEvent) {
4.         int metaState = keyEvent.getMetaState();
5.         int unicodeChar = keyEvent.getUnicodeChar();
6.         String msg = "";
7.         msg +="按键动作:" + String.valueOf(keyEvent.getAction())+"\n";
8.         msg +="按键代码:" + String.valueOf(keyCode)+"\n";
9.         msg +="按键字符:" + (char)unicodeChar+"\n";
10.        msg +="UNICODE:" + String.valueOf(unicodeChar)+"\n";
11.        msg +="重复次数:" + String.valueOf(keyEvent.getRepeatCount())+"\n";
12.        msg +="功能键状态:" + String.valueOf(metaState)+"\n";
13.        msg +="硬件编码:" + String.valueOf(keyEvent.getScanCode())+"\n";
14.        msg +="按键标志:" + String.valueOf(keyEvent.getFlags())+"\n";
15.        labelView.setText(msg);
```

```
16.     if (checkBox.isChecked())
17.         return true;
18.     else
19.         return false;
20. }
```

如上所示,第 4 行代码用来获取功能键状态。功能键包括左 Alt 键、右 Alt 键和 Shift 键,当这三个功能键被按下时,功能键代码 metaState 值分别为 18、34 和 65;但没有功能键被按下时,功能键代码 metaState 值为 0。第 5 行代码获取了按键的 Unicode 值,在第 9 行中,将 Unicode 转换为字符,显示在 TextView 中。第 7 行代码获取了按键动作,0 表示按下按键,1 表示抬起按键。第 7 行代码获取按键的重复次数,但按键被长时间按下时,则会产生这个属性值。

第 13 行代码获取了按键的硬件编码,不同硬件设备的按键硬件编码都不相同,因此该值一般用于调试。第 14 行获取了按键事件的标志符。

5.5.2 触摸事件

Android 界面框架支持对触摸事件的监听,并能够将触摸事件的详细信息传递给处理函数,需要设置触摸事件的监听器,并重载 onTouch()函数。

```
1. touchView.setOnTouchListener(new View.OnTouchListener(){
2. @Override
3. public boolean onTouch(View v, MotionEvent event) {
4.     //过程代码……
5.     return true/false;
6. }
```

如上所示,第 1 行代码是设置控件的触摸事件监听器。在代码第 3 行的 onTouch()函数中,第 1 个参数 View 表示产生触摸事件的界面控件;第 2 个参数 MontionEvent 表示触摸事件的详细信息,如产生时间、坐标和触点压力等,第 5 行是 onTouch()函数的返回值。

TouchEventDemo 是一个说明如何处理触摸事件的示例。如图 5-32 所示是一个 TouchEventDemo 用户界面,浅蓝色区域是可以接受触摸事件的区域,用户可以在 Android 模拟器中使用鼠标点击屏幕,用以模拟触摸手机屏幕,下方黑色区域是显示区域,用来显示触摸事件的类型、相对坐标、绝对坐标、触点压力、触点尺寸和历史数据量等信息。

在用户界面中使用了线性布局,并加入了 3 个 TextView 控件。第 1 个 TextView(ID 为 touch_area)用来标识触摸事件的测试区域,第 2 个 TextView(ID 为 history_label)用来显示触摸事件的历史数据量,第 3 个

图 5-32 "TouchEventDemo"视图效果

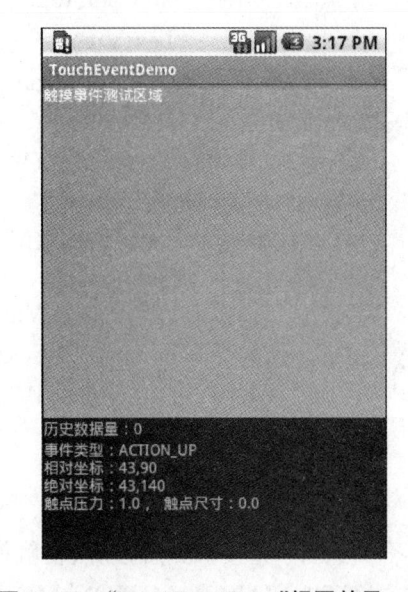

TextView(ID 为 event_label)用来表示触摸事件的详细信息,包括类型、相对坐标、绝对坐标、触点压力和触点尺寸。

XML 文件的代码如下。

```
1. <?xml version="1.0" encoding="utf-8"?>
2. <LinearLayout xmlns:android="http://schemas.android.com/apk/res/android"
3.    android:orientation="vertical"
4.    android:layout_width="fill_parent"
5.    android:layout_height="fill_parent">
6. <TextView   android:id="@+id/touch_area"
7.      android:layout_width="fill_parent"
8.      android:layout_height="300dip"
9.      android:background="#0FF"
10.     android:textColor="#FFFFFF"
11.     android:text="触摸事件测试区域">
12. </TextView>
13. <TextView android:id="@+id/history_label"
14.     android:layout_width="wrap_content"
15.     android:layout_height="wrap_content"
16.     android:text="历史数据量:">
17. </TextView>
18. <TextView android:id="@+id/event_label"
19.     android:layout_width="wrap_content"
20.     android:layout_height="wrap_content"
21.     android:text="触摸事件:">
22. </TextView>
23.  </LinearLayout>
```

第 9 行代码定义了 TextView 的背景颜色,♯80A0FF 是颜色代码。第 10 行代码定义了 TextView 的字体颜色。

在代码中为了能够引用 XML 文件中声明的界面元素,使用了下面的代码。

```
1. TextView labelView = null;
2. labelView = (TextView)findViewById(R.id.event_label);
3. TextView touchView = (TextView)findViewById(R.id.touch_area);
4. final TextView historyView = (TextView)findViewById(R.id.history_label);
```

当手指接触到触摸屏、在触摸屏上移动或离开触摸屏时,分别会引发 ACTION_DOWN、ACTION_UP 和 ACTION_MOVE 触摸事件,而无论是哪种触摸事件,都会调用 onTouch()

函数进行处理。事件类型包含在 onTouch()函数的 MotionEvent 参数中,可以通过 getAction()函数获取到触摸事件的类型,然后根据触摸事件的不同类型进行不同的处理。为了能够使屏幕最上方的 TextView 处理触摸事件,需要使用 setOnTouchListener()函数在代码中设置触摸事件监听器,并在 onTouch()函数添加触摸事件的处理过程。

```
1.  touchView.setOnTouchListener(new View.OnTouchListener(){
2.    @Override
3.    public boolean onTouch(View v, MotionEvent event) {
4.        int action = event.getAction();
5.        switch (action) {
6.          case (MotionEvent.ACTION_DOWN):
7.            Display("ACTION_DOWN",event);
8.            break;
9.          case (MotionEvent.ACTION_UP):
10.           int historySize = ProcessHistory(event);
11.           historyView.setText("历史数据量:"+historySize);
12.           Display("ACTION_UP",event);
13.           break;
14.         case (MotionEvent.ACTION_MOVE):
15.           Display("ACTION_MOVE",event);
16.           break;
17.       }
18.       return true;
19.    }
20. });
```

第 7 行代码的 Display()是一个自定义函数,主要用来显示触摸事件的详细信息,函数的代码和含义将在后面进行介绍。第 10 行代码的 ProcessHistory()也是一个自定义函数,用来处理触摸事件的历史数据,后面进行介绍。第 11 行代码使用 TextView 显示历史数据的数量。

MotionEvent 参数中不仅有触摸事件的类型信息,还包括触点的坐标信息,获取方法是使用 getX()和 getY()函数。这两个函数获取到的是触点相对于父界面元素的坐标信息。如果需要获取绝对坐标信息,则可使用 getRawX()和 getRawY()函数。触点压力是一个介于 0 和 1 之间的浮点数,用来表示用户对触摸屏施加压力的大小,接近 0 表示压力较小,接近 1 表示压力较大。获取触摸事件触点压力的方式是调用 getPressure()函数。触点尺寸指用户接触触摸屏的接触点大小,也是一个介于 0 和 1 之间的浮点数,接近 0 表示尺寸较小,接近 1 表示尺寸较大,可以使用 getSize()函数获取。

Display()将 MotionEvent 参数参数中的事件信息提取出来,并显示在用户界面上。

```
1.  private void Display(String eventType, MotionEvent event){
2.    int x = (int)event.getX();
3.    int y = (int)event.getY();
```

```
4.    float pressure = event.getPressure();
5.    float size = event.getSize();
6.    int RawX = (int)event.getRawX();
7.    int RawY = (int)event.getRawY();
8.
9.    String msg = "";
10.   msg += "事件类型:" + eventType + "\n";
11.   msg += "相对坐标:"+String.valueOf(x)+","+String.valueOf(y)+"\n";
12.   msg += "绝对坐标:"+String.valueOf(RawX)+","+String.valueOf(RawY)+"\n";
13.   msg += "触点压力:"+String.valueOf(pressure)+",  ";
14.   msg += "触点尺寸:"+String.valueOf(size)+"\n";
15.   labelView.setText(msg);
16. }
```

一般情况下,如果用户将手指按在触摸屏上,但不移动,然后抬起手指,应先后产生 ACTION_DOWN 和 ACTION_UP 两个触摸事件。如果用户在屏幕上移动手指,然后再抬起手指,则会产生这样的事件序列:ACTION_DOWN → ACTION_MOVE → ACTION_MOVE → ACTION_MOVE → ……→ ACTION_UP。

在手机上运行的应用程序,效率是非常重要的。如果 Android 界面框架不能产生足够多的触摸事件,则应用程序就不能够很精确的描绘触摸屏上的触摸轨迹;如果 Android 界面框架产生了过多的触摸事件,虽然能够满足精度的要求,但却降低了应用程序效率。Android 界面框架使用了"打包"的解决方法。在触点移动时,每经过一定的时间间隔便会产生一个 ACTION_MOVE 事件,在这个事件中,除了有当前触点的相关信息外,还包含这段时间间隔内触点轨迹的历史数据信息。这样既能够保持精度,又不至于产生过多的触摸事件。

通常情况下,在 ACTION_MOVE 的事件处理函数中,都先处理历史数据,然后再处理当前数据。

```
1. private int ProcessHistory(MotionEvent event)
2. {
3.    int historySize = event.getHistorySize();
4.    for (int i = 0; i < historySize; i++) {
5.        long time = event.getHistoricalEventTime(i);
6.        float pressure = event.getHistoricalPressure(i);
7.        float x = event.getHistoricalX(i);
8.        float y = event.getHistoricalY(i);
9.        float size = event.getHistoricalSize(i);
10.
11.       // 处理过程……
12.   }
```

```
13.     return historySize;
14. }
```

如上所示,第 3 行代码获取了历史数据的数量,然后在第 4 行至 12 行中循环处理这些历史数据。第 5 行代码获取了历史事件的发生时间。第 6 行代码获取历史事件的触点压力。第 7 行和第 8 行代码获取历史事件的相对坐标。第 9 行获取历史事件的触点尺寸,在第 14 行返回历史数据的数量,主要是用于界面显示。Android 模拟器并不支持触点压力和触点尺寸的模拟,所有触点压力恒为 1.0,触点尺寸恒为 0.0,同时 Android 模拟器上也无法产生历史数据,因此历史数据量一直显示为 0。

习题与思考题

1. 简述五种界面布局的特点。

2. 参考下图中界面控件的摆放位置,分别使用线性布局、相对布局和绝对布局实现用户界面,并对比各种布局实现的复杂程度和对不同屏幕尺寸的适应能力。

3. 简述 Android 系统支持的三种菜单。

4. EditText 控件有 Numeric 属性,设置成 integer 后 EditText 控件中只能输入数字,无法输入其他字母或符号。利用按键事件,编程实现 EditText 控件的这一功能。

扩展练习

制作一个有自动提示功能的应用程序。

第6章 组件通信与广播消息

☆6.1 Intent
☆6.2 广播消息

本章学习目标：了解使用 Intent 进行组件通信的原理；掌握使用 Intent 启动 Activity 的方法，掌握获取 Activity 返回值的方法；了解 Intent 过滤器的原理与匹配机制，掌握发送和接收广播消息的方法。

6.1 Intent

Intent 是一个动作的完整描述，包含了动作的产生组件、接收组件和传递的数据信息。Intent 也可称为一个在不同组件之间传递的消息。这个消息在到达接收组件后，接收组件会执行相关的动作。Intent 为 Activity、Service 和 BroadcastReceiver 等组件提供交互能力。

Intent 的用途主要是启动 Activity 和 Service，在 Android 系统上发布广播消息，广播消息可以是接收到特定数据或消息，也可以是手机的信号变化或电池的电量过低等信息。

6.1.1 启动 Activity

在 Android 系统中，应用程序一般都有多个 Activity，Intent 可以实现不同 Activity 之间的切换和数据传递。

启动 Activity 方式有两种：显式启动，必须在 Intent 中指明启动的 Activity 所在的类；隐式启动，Android 系统根据 Intent 的动作和数据来决定启动哪一个 Activity，也就是说 Intent 中只包含需要执行的动作和所包含的数据，而无需指明具体启动哪一个 Activity，选择权由 Android 系统和最终用户来决定。使用 Intent 显式启动 Activity，先创建一个 Intent，指定当前的应用程序上下文以及要启动的 Activity，把创建好的这个 Intent 作为参数传递给 startActivity()即可。

```
1. Intent intent = new Intent(IntentDemo.this, ActivityToStart.class);
2. startActivity(intent);
```

IntentDemo 示例说明如何使用 Intent 启动新的 Activity，IntentDemo 示例包含 IntentDemo、ActivityToStart 这两个 Activity 类，程序默认启动 IntentDemo 这个 Activity。

图 6-1　IntentDemo 示例　　　　图 6-2　启动 Activity 后的界面

如图 6-2 所示，点击"启动 Activity"按钮后，程序启动 ActivityToStart 这个 Activity。

在 AndroidManifest.xml 文件中注册上面这两个 Activity，应使用＜activity＞标签，嵌套在＜application＞标签内部。

```xml
1. <?xml version="1.0" encoding="utf-8"?>
2. <manifest xmlns:android="http://schemas.android.com/apk/res/android"
3.           package="edu.xzceu.IntentDemo"
4.           android:versionCode="1"
5.           android:versionName="1.0">
6.     <application android:icon="@drawable/icon"
           android:label="@string/app_name">
7.         <activity android:name=".IntentDemo"
8.                   android:label="@string/app_name">
9.             <intent-filter>
10.                 <action android:name="android.intent.action.MAIN" />
11.                 <category android:name="android.intent.category.LAUNCHER" />
12.             </intent-filter>
13.         </activity>
14.         <activity android:name=".ActivityToStart"
15.                   android:label="@string/app_name">
16.         </activity>
17.     </application>
18.     <uses-sdk android:minSdkVersion="3" />
19. </manifest>
```

Android 应用程序中,用户使用的每个组件都必须在 AndroidManifest.xml 文件中的<application>节点内定义,<application>节点下共有两个<activity>节点,分别代表应用程序中所使用的两个 Activity,IntentDemo 和 ActivityToStart。

在 IntentDemo.java 文件中,包含了显示使用 Intent 启动 Activity 的核心代码。

```java
1. Button button = (Button)findViewById(R.id.btn);
2. button.setOnClickListener(new OnClickListener(){
3.     public void onClick(View view){
4.         Intent intent = new Intent(IntentDemo.this, ActivityToStart.class);
5.         startActivity(intent);
6.     }
7. });
```

在点击事件的处理函数中,Intent 构造函数的第 1 个参数是应用程序上下文,程序中的应用程序上下文就是 IntentDemo;第 2 个参数是接收 Intent 的目标组件,使用的是显式启动方式,直接指明了需要启动的 Activity。

隐式启动的优点是不需要指明需要启动哪一个 Activity,而由 Android 系统来决定,有利于使用第三方组件。隐式启动 Activity 时,Android 系统在应用程序运行时解析 Intent,并根据一定的规则对 Intent 和 Activity 进行匹配,使 Intent 上的动作、数据与 Activity 完全吻合。

匹配的 Activity 可以是应用程序本身的,也可以是 Android 系统内置的,还可以是第三方应用程序提供的。因此,这种方式更加强调了 Android 应用程序中组件的可复用性。

在缺省情况下,Android 系统会调用内置的 Web 浏览器。

1. Intent intent = new Intent(Intent. ACTION_VIEW, Uri. parse("http://www.google.com"));
2. startActivity(intent);

Intent 的动作是 Intent. ACTION_VIEW,根据 URI 的数据类型来匹配动作。数据部分的 URI 是 Web 地址,使用 Uri. parse(urlString)方法,可以简单的把一个字符串解释成 Uri 对象,Intent 的语法如下:

Intent intent = new Intent(Intent. ACTION_VIEW, Uri. parse(urlString));

Intent 构造函数的第 1 个参数是 Intent 需要执行的动作,第 2 个参数是 URI,表示需要传递的数据。Android 系统支持的常见动作字符串常量表,如下表 6-1 所示。

表 6-1 Android 系统支持的常见动作字符串常量表

动 作	说 明
ACTION_ANSWER	打开接听电话的 Activity,默认为 Android 内置的拨号盘界面
ACTION_CALL	打开拨号盘界面并拨打电话,使用 Uri 中的数字部分作为电话号码
ACTION_DELETE	打开一个 Activity,对所提供的数据进行删除操作
ACTION_DIAL	打开内置拨号盘界面,显示 Uri 中提供的电话号码
ACTION_EDIT	打开一个 Activity,对所提供的数据进行编辑操作
ACTION_INSERT	打开一个 Activity,在提供数据的当前位置插入新项
ACTION_PICK	启动一个子 Activity,从提供的数据列表中选取一项
ACTION_SEARCH	启动一个 Activity,执行搜索动作
ACTION_SENDTO	启动一个 Activity,向数据提供的联系人发送信息
ACTION_SEND	启动一个可以发送数据的 Activity
ACTION_VIEW	最常用的动作,对以 Uri 方式传送的数据,根据 Uri 协议部分以最佳方式启动相应的 Activity 进行处理。对于 http:address 将打开浏览器查看;对于 tel:address 将打开拨号呼叫指定的电话号码
ACTION_WEB_SEARCH	打开一个 Activity,对提供的数据进行 Web 搜索

WebViewIntentDemo 示例说明如何隐式启动 Activity。

如图 6-3 所示,当用户在文本框中输入要访问网址后,通过点击"浏览此 URL"按钮,程序根据用户输入的网址生成一个 Intent,并以隐式启动的方式调用 Android 内置的 Web 浏览器,打开指定的 Web 页面。本例输入的网址是南京晓庄学院的主站地址,地址是:http://

www.njxzc.edu.cn/，如图 6-4 所示。

图 6-3 打开指定的 Web 页面

图 6-4 南京晓庄学院的主站页面

6.1.2 获取 Activity 返回值

在上一小节 IntentDemo 示例中，通过使用 startActivity(Intent)方法启动 Activity 后，两个 Activity 之间相互独立，没有任何的关联。在很多情况下，后启动的 Activity 是为了让用户对特定信息进行选择，在关闭这个 Activity 后，用户的选择信息需要返回给未关闭的那个 Activity。按照 Activity 启动的先后顺序，先启动的称为父 Activity，后启动的称为子 Activity。

如果需要将子 Activity 的部分信息返回给父 Activity，则可以使用 Sub-Activity 的方式去启动子 Activity，获取子 Activity 的返回值。一般可以分为以下三个步骤：以 Sub-Activity 的方式启动子 Activity，设置子 Activity 的返回值，在父 Activity 中获取返回值。

以 Sub-Activity 的方式启动子 Activity，调用 startActivityForResult（Intent，requestCode）函数，参数 Intent 用于决定启动哪个 Activity。参数 requestCode 是唯一的标识子 Activity 的请求码，显式启动子 Activity 的代码如下：

```
1. int SUBACTIVITY1 = 1;
2. Intent intent = new Intent(this, SubActivity1.class);
3. startActivityForResult(intent, SUBACTIVITY1);
```

隐式启动子 Activity 的代码如下。

```
1. int SUBACTIVITY2 = 2;
2. Uri uri = Uri.parse("content://contacts/people");
3. Intent intent = new Intent(Intent.ACTION_PICK, uri);
4. startActivityForResult(intent, SUBACTIVITY2);
```

设置子 Activity 的返回值,需在子 Activity 调用 finish()函数关闭前,调用 setResult()函数将所需的数据返回给父 Activity。setResult()函数有两个参数:结果码、返回值。结果码表明了子 Activity 的返回状态,通常为 Activity.RESULT_OK 或者 Activity.RESULT_CANCELED,或自定义的结果码,结果码均为整数类型。返回值用封装在 Intent 中,子 Activity 通过 Intent 将需要返回的数据传递给父 Activity。数据主要是 Uri 形式,可以附加一些额外信息,这些额外信息用 Extra 的集合表示。

下面代码说明如何在子 Activity 中设置返回值。

```
1. Uri data = Uri.parse("tel:" + tel_number);
2. Intent result = new Intent(null, data);
3. result.putExtra("address", " ");
4. setResult(RESULT_OK, result);
5. finish();
```

在父 Activity 中获取返回值,当子 Activity 关闭时,启动它的父 Activity 的 onActivityResult()函数将被调用;如果需要在父 Activity 中处理子 Activity 的返回值,则重载此函数即可,此函数的语法如下。

```
1. public void onActivityResult(int requestCode, int resultCode, Intent data);
```

第 1 个参数 requestCode,用来表示是哪一个子 Activity 的返回值;第 2 个参数 resultCode 用于表示子 Activity 的返回状态;第 3 个参数 data 是子 Activity 的返回数据,返回数据类型是 Intent。根据返回数据的用途不同,Uri 数据的协议则不同,也可以使用 Extra 方法返回一些原始类型的数据。

下面代码说明如何在父 Activity 中处理子 Activity 的返回值。

```
1. private static final int SUBACTIVITY1 = 1;
2. private static final int SUBACTIVITY2 = 2;
3.
4. @Override
```

```
5. public void onActivityResult(int requestCode, int resultCode, Intent data){
6.     Super.onActivityResult(requestCode, resultCode, data);
7.     switch(requestCode){
8.         case SUBACTIVITY1:
9.             if (resultCode == Activity.RESULT_OK){
10.                Uri uriData = data.getData();
11.            }else if (resultCode == Activity.RESULT_CANCEL){
12.            }
13.            break;
14.        case SUBACTIVITY2:
15.            if (resultCode == Activity.RESULT_OK){
16.                Uri uriData = data.getData();
17.            }
18.            break;
19.     }
20. }
```

第 1 行代码和第 12 行代码是两个子 Activity 的请求码,第 7 行代码对请求码进行匹配。第 9 行和第 11 行代码对结果码进行判断:如果返 4 回的结果码是 Activity.RESULT_OK,则在代码的第 10 行使用 getData()函数获取 Intent 中的 Uri 数据;如果返回的结果码是 Activity.RESULT_CANCELED,则不进行任何操作。ActivityCommunication 示例说明了如何以 Sub-Activity 方式启动子 Activity,以及使用 Intent 进行组件间通信。如图 6-5 所示。

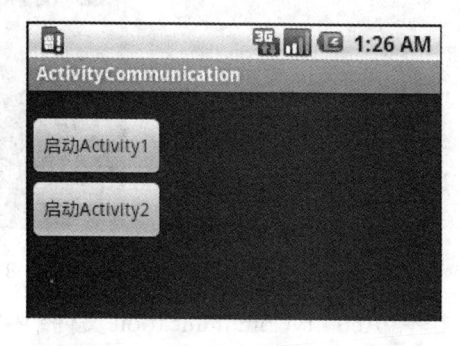

图 6-5 ActivityCommunication 示例

当用户点击"启动 Activity1"和"启动 Activity2"按钮时,如图 6-6 所示,程序将分别启动子 SubActivity1 和 SubActivity2。

图 6-6 程序启动子 SubActivity1 和 SubActivity2

如图 6-7 所示,SubActivity1 提供了一个输入框,以及"接受"和"撤销"两个按钮。如果在输入框中输入信息后点击"接受"按钮,程序会把输入框中的信息传递给其父 Activity,并在父 Activity 的界面上显示；如果用户点击"撤销"按钮，则程序不会向父 Activity 传递任何信息。

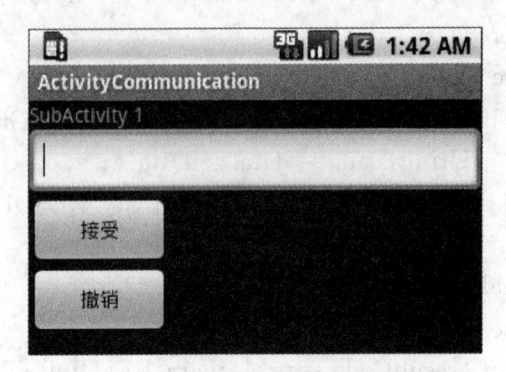

图 6-7 SubActivity1 提供的输入框

SubActivity2 主要是为了说明如何在父 Activity 中处理多个子 Activity，因此仅提供了用于关闭 SubActivity2 的"关闭"按钮，如图 6-8 所示。

图 6-8 SubActivity2 提供的输入框

ActivityCommunication 文件结构，如图 6-9 所示。

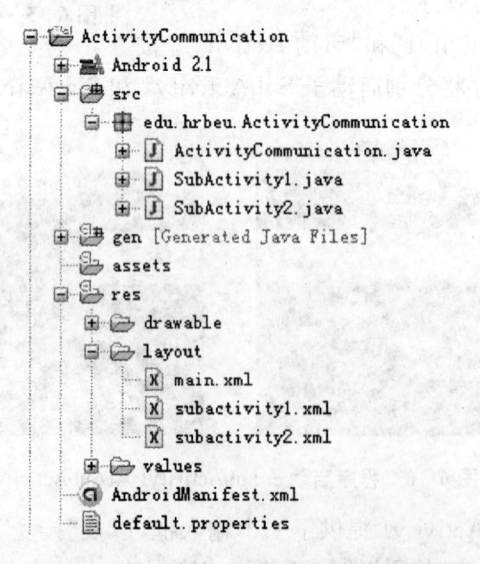

图 6-9 ActivityCommunication 文件结构

ActivityCommunication.java 文件的核心代码如下。

```
1.  public class ActivityCommunication extends Activity {
2.      private static final int SUBACTIVITY1 = 1;
3.      private static final int SUBACTIVITY2 = 2;
4.      TextView textView;
5.      @Override
6.      public void onCreate(Bundle savedInstanceState) {
7.          super.onCreate(savedInstanceState);
8.          setContentView(R.layout.main);
9.          textView = (TextView)findViewById(R.id.textShow);
10.         final Button btn1 = (Button)findViewById(R.id.btn1);
11.         final Button btn2 = (Button)findViewById(R.id.btn2);
12.
13.         btn1.setOnClickListener(new OnClickListener(){
14.             public void onClick(View view){
15.                 Intent intent = new Intent(ActivityCommunication.this, SubActivity1.class);
16.                 startActivityForResult(intent, SUBACTIVITY1);
17.             }
18.         });
19.
20.         btn2.setOnClickListener(new OnClickListener(){
21.             public void onClick(View view){
22.                 Intent intent = new Intent(ActivityCommunication.this, SubActivity2.class);
23.                 startActivityForResult(intent, SUBACTIVITY2);
24.             }
25.         });
26.     }
27.
28.     @Override
29.     protected void onActivityResult(int requestCode, int resultCode, Intent data) {
30.         super.onActivityResult(requestCode, resultCode, data);
31.
32.         switch(requestCode){
33.         case SUBACTIVITY1:
34.             if (resultCode == RESULT_OK){
```

```
35.            Uri uriData = data.getData();
36.            textView.setText(uriData.toString());
37.        }
38.        break;
39.    case SUBACTIVITY2：
40.        break;
41.    }
42.  }
43. }
```

代码的第 2 行和第 3 行分别定义了两个子 Activity 的请求码，在代码的第 16 行和第 23 行以 Sub-Activity 的方式分别启动两个子 Activity。代码第 29 行是子 Activity 关闭后的返回值处理函数，其中 requestCode 是子 Activity 返回的请求码，应该与第 2 行和第 3 行定义的两个请求码相匹配；resultCode 是结果码，在代码第 32 行对结果码进行判断，如果等于 RESULT_OK，在第 35 行代码获取子 Activity 的返回值中的数据。data 是返回值，子 Activity 需要返回的数据就保存在 data 中。

SubActivity1.java 的核心代码如下。

```
1.  public class SubActivity1 extends Activity {
2.    @Override
3.    public void onCreate(Bundle savedInstanceState) {
4.        super.onCreate(savedInstanceState);
5.        setContentView(R.layout.subactivity1);
6.        final EditText editText = (EditText)findViewById(R.id.edit);
7.        Button btnOK = (Button)findViewById(R.id.btn_ok);
8.        Button btnCancel = (Button)findViewById(R.id.btn_cancel);
9.
10.       btnOK.setOnClickListener(new OnClickListener(){
11.           public void onClick(View view){
12.               String uriString = editText.getText().toString();
13.               Uri data = Uri.parse(uriString);
14.               Intent result = new Intent(null, data);
15.               setResult(RESULT_OK, result);
16.               finish();
17.           }
18.       });
19.
20.       btnCancel.setOnClickListener(new OnClickListener(){
21.           public void onClick(View view){
22.               setResult(RESULT_CANCELED, null);
```

```
23.            finish();
24.         }
25.     });
26.   }
27. }
```

第 13 行代码将 EditText 控件的内容作为数据保存在 Uri 中。第 14 行代码中使用这个 Uri 构造 Intent。第 15 行代码中，将 Intent 作为返回值，RESUIT_OK 作为结果码，通过调用 setResult()函数，将返回值和结果码传递给父 Activity。第 16 行代码调用 finish()函数关闭当前的子 Activity。SubActivity2.java 的核心代码如下。

```
1.  public class SubActivity2 extends Activity {
2.      @Override
3.      public void onCreate(Bundle savedInstanceState) {
4.          super.onCreate(savedInstanceState);
5.          setContentView(R.layout.subactivity2);
6.
7.          Button btnReturn = (Button)findViewById(R.id.btn_return);
8.          btnReturn.setOnClickListener(new OnClickListener(){
9.              public void onClick(View view){
10.                 setResult(RESULT_CANCELED, null);
11.                 finish();
12.             }
13.         });
14.     }
15. }
```

第 10 行的 setResult()函数仅设置了结果码，第 2 个参数为 null，表示数据需要传递给父 Activity。

6.1.3 Intent 过滤器

Intent 过滤器是一种根据 Intent 中的动作（Action）、类别（Categorie）和数据（Data）等内容，对适合接收该 Intent 的组件进行匹配和筛选的机制，Intent 过滤器可以匹配数据类型、路径和协议，还包括可以用来确定多个匹配项顺序的优先级（Priority）。应用程序的 Activity 组件、Service 组件和 BroadcastReceiver 都可以注册 Intent 过滤器，则这些组件在特定的数据格式上就可以产生相应的动作。

1. 注册 Intent 过滤器

在 AndroidManifest.xml 文件的各个组件的节点下定义<intent-filter>节点，然后在<intent-filter>节点中声明该组件所支持的动作、执行的环境和数据格式等信息，在程序代码中动态地为组件设置 Intent 过滤器。<intent-filter>节点支持<action>标签、<category>标签和<data>标签，<action>标签定义 Intent 过滤器的"类别"，<category>标签定义 Intent

过滤器的"动作",<data>标签定义 Intent 过滤器的"数据"。

<intent-filter>节点支持的标签和属性,如下表 6-2 所示。

表 6-2 <intent-filter>节点支持的标签和属性

标 签	属 性	说 明
<action>	android:name	指定组件所能响应的动作,用字符串表示,通常使用 Java 类名和包的完全限定名构成
<category>	android:category	指定以何种方式去服务 Intent 请求的动作
<data>	Android:host	指定一个有效的主机名
	android:mimetype	指定组件能处理的数据类型
	android:path	有效的 URI 路径名
	android:port	主机的有效端口号
	android:scheme	所需要的特定协议

<category>标签用来指定 Intent 过滤器的服务方式,每个 Intent 过滤器可以定义多个<category>标签,程序开发人员可使用自定义的类别,或使用 Android 系统提供的类别。

Android 系统提供的类别,如下表 6-3 所示。

表 6-3 Android 系统提供的类别

值	说 明
ALTERNATIVE	Intent 数据默认动作的一个可替换的执行方法
SELECTED_ALTERNATIVE	和 ALTERNATIVE 类似,但替换的执行方法不是指定的,而是被解析出来的
BROWSABLE	声明 Activity 可以由浏览器启动
DEFAULT	为 Intent 过滤器中定义的数据提供默认动作
HOME	设备启动后显示的第一个 Activity
LAUNCHER	在应用程序启动时首先被显示

AndroidManifest.xml 文件中的每个组件的<intent-filter>都被解析成一个 Intent 过滤器对象。当应用程序安装到 Android 系统时,所有的组件和 Intent 过滤器都会注册到 Android 系统中。这样,Android 系统便知道了如何将任意一个 Intent 请求通过 Intent 过滤器映射到相应的组件上。

2. Intent 解析

Intent 到 Intent 过滤器的映射过程称为"Intent 解析"。Intent 解析可以在所有的组件中,找到一个可以与请求的 Intent 达成最佳匹配的 Intent 过滤器。

Android 系统把所有应用程序包中的 Intent 过滤器集合在一起,形成一个完整的 Intent 过滤器列表,在 Intent 与 Intent 过滤器进行匹配时,Android 系统会将列表中所有 Intent 过滤器的"动作"和"类别"与 Intent 进行匹配,任何不匹配的 Intent 过滤器都将被过滤掉。没有指定"action"的 Intent 过滤器可以匹配任何的 Intent,但是没有指定"category"的 Intent 过滤器只能匹配没有"category"的 Intent。把 Intent 数据 Uri 的每个子部与 Intent 过滤器的<data

>标签中的属性进行匹配,如果<data>标签指定了协议、主机名、路径名或 MIME 类型,那么这些属性都要与 Intent 的 Uri 数据部分进行匹配,任何不匹配的 Intent 过滤器均被过滤掉。如果 Intent 过滤器的匹配结果多于一个,则可以根据在<intent-filter>标签中定义的优先级标签来对 Intent 过滤器进行排序,优先级最高的 Intent 过滤器将被选择。

IntentResolutionDemo 示例说明了如何在 AndroidManifest.xml 文件中注册 Intent 过滤器,以及如何设置<intent-filter>节点属性来捕获指定的 Intent。

AndroidManifest.xml 的完整代码如下。

```
1.  <? xml version="1.0" encoding="utf-8"?>
2.  <manifest xmlns:android="http://schemas.android.com/apk/res/android"
3.      package="edu.xzceu.IntentResolutionDemo"
4.      android:versionCode="1"
5.      android:versionName="1.0">
6.      <application android:icon="@drawable/icon"
            android:label="@string/app_name">
7.          <activity android:name=".IntentResolutionDemo"
8.              android:label="@string/app_name">
9.              <intent-filter>
10.                 <action android:name="android.intent.action.MAIN" />
11.                 <category android:name="android.intent.category.LAUNCHER" />
12.             </intent-filter>
13.         </activity>
14.         <activity android:name=".ActivityToStart"
15.             android:label="@string/app_name">
16.             <intent-filter>
17.                 <action android:name="android.intent.action.VIEW" />
18.                 <category android:name="android.intent.category.DEFAULT" />
19.                 <data android:scheme="schemodemo" android:host="edu.xzceu" />
20.             </intent-filter>
21.         </activity>
22.     </application>
23.     <uses-sdk android:minSdkVersion="3" />
24. </manifest>
```

第 7 行代码和第 14 行代码分别定义了两个 Activity。第 9 行到第 12 行是第 1 个 Activity 的 Intent 过滤器,动作是 android.intent.action.MAIN,类别是 android.intent.category.LAUNCHER。由过滤器的动作和类别可知,这个 Activity 是应用程序启动后显示的缺省用户界面。第 16~20 行是第 2 个 Activity 的 Intent 过滤器,过滤器的动作是 android.

intent.action.VIEW,表示根据 Uri 协议,以最佳的方式启动相应的 Activity;类别是 android.intent.category.DEFAULT,表示数据的默认动作;数据的协议部分是 android:scheme="schemodemo",数据的主机名称部分是 android:host="edu.xzceu" schemodemo://edu.xzceu。

IntentResolutionDemo.java 文件中 Intent 实例化和启动 Activity 的代码如下。

```
1. Intent intent = new Intent(Intent.ACTION_VIEW,
      Uri.parse("schemodemo://edu.xzceu/path"));
2. startActivity(intent);
```

第 1 行代码定义的 Intent,动作为 Intent.ACTION_VIEW,与 Intent 过滤器的动作 android.intent.action.VIEW 匹配;Uri 是"schemodemo://edu.xzceu/path",其中的协议部分为"schemodemo",主机名部分为"edu.xzceu",也与 Intent 过滤器定义的数据要求完全匹配。代码第 1 行定义的 Intent,在 Android 系统与 Intent 过滤器列表进行匹配时,会与 AndroidManifest.xml 文件中 ActivityToStart 定义的 Intent 过滤器完全匹配。

6.2 广播消息

Intent 的另一种用途是发送广播消息,应用程序和 Android 系统都可以使用 Intent 发送广播消息。广播消息的内容可以是与应用程序密切相关的数据信息,也可以是 Android 的系统信息,例如网络连接变化、电池电量变化、接收到短信和系统设置变化等等。如果应用程序注册了 BroadcastReceiver,则可以接收到指定的广播消息。

广播信息的使用方法,创建一个 Intent,通常使用应用程序包的名称,调用 sendBroadcast()函数,就可把 Intent 携带的消息广播出去。注意:在构造 Intent 时必须用一个全局唯一的字符串标识其要执行的动作。如果要在 Intent 中传递额外数据,可以用 Intent 的 putExtra()方法,利用 Intent 发送广播消息,并添加了额外的数据,然后调用 sendBroadcast()发送了广播消息的代码。

```
1. String UNIQUE_STRING = "edu.xzceu.BroadcastReceiverDemo";
2. Intent intent = new Intent(UNIQUE_STRING);
3. intent.putExtra("key1", "value1");
4. intent.putExtra("key2", "value2");
5. sendBroadcast(intent);
```

第 1 行代码中的 BroadcastReceiver 用于监听广播消息,可以在 AndroidManifest.xml 文件或在代码中注册一个 BroadcastReceiver,并使用 Intent 过滤器指定要处理的广播消息。创建 BroadcastReceiver 需继承 BroadcastReceiver 类,并重载 onReceive()方法。代码如下。

```
1. public class MyBroadcastReceiver extends BroadcastReceiver {
2. @Override
3.     public void onReceive(Context context, Intent intent) {
4.         //TODO: React to the Intent received.
5.     }
6. }
```

BroadcastReceiver 的应用程序不需要一直运行,当 Android 系统接收到与之匹配的广播消息时,会自动启动此 BroadcastReceiver。基于以上的特征,BroadcastReceiver 适合做一些资源管理的工作。在 BroadcastReceiver 接收到与之匹配的广播消息后,onReceive()方法会被调用且必须要在 5 秒钟执行完毕,否则 Android 系统会认为该组件失去响应,并提示用户强行关闭该组件。如图 6-10,BroadcastReceiverDemo 示例说明了如何在应用程序中注册 BroadcastReceiver,并接收指定类型的广播消息。在点击"发生广播消息"按钮后,EditText 控件中内容将以广播消息的形式发送出去,示例内部的 BroadcastReceiver 将接收这个广播消息,并显示在用户界面的下方。

图 6-10 BroadcastReceiverDemo 示例

BroadcastReceiverDemo.java 文件中包含发送广播消息的代码,其关键代码如下。

```
1. button.setOnClickListener(new OnClickListener(){
2.     public void onClick(View view){
3.         Intent intent = new Intent("edu.xzceu.BroadcastReceiverDemo");
4.         intent.putExtra("message", entryText.getText().toString());
5.         sendBroadcast(intent);
6.     }
7. });
```

第 3 行代码创建 Intent，将 edu.xzceu.BroadcastReceiverDem 作为识别广播消息的字符串标识。第 4 行代码添加了额外信息。第 5 行代码调用 sendBroadcast()函数发送广播消息。为了能够使应用程序中的 BroadcastReceiver 接收指定的广播消息，首先要在 AndroidManifest.xml 文件中添加 Intent 过滤器，声明 BroadcastReceiver 可以接收的广播消息。

AndroidManifest.xml 文件的完整代码如下。

```
1. <?xml version="1.0" encoding="utf-8"?>
2. <manifest xmlns:android="http://schemas.android.com/apk/res/android"
3.     package="edu.xzceu.BroadcastReceiverDemo"
4.     android:versionCode="1"
5.     android:versionName="1.0">
6.     <application android:icon="@drawable/icon" android:label="@string/app_name">
7.         <activity android:name=".BroadcastReceiverDemo"
8.                   android:label="@string/app_name">
9.             <intent-filter>
10.                 <action android:name="android.intent.action.MAIN" />
11.                 <category android:name="android.intent.category.LAUNCHER" />
12.             </intent-filter>
13.         </activity>
14.         <receiver android:name=".MyBroadcastReceiver">
15.             <intent-filter>
16.                 <action android:name="edu.xzceu.BroadcastReceiverDemo" />
17.             </intent-filter>
18.         </receiver>
19.     </application>
20.     <uses-sdk android:minSdkVersion="3" />
21. </manifest>
```

第 14 行代码中创建了一个<receiver>节点，在第 15 行中声明了 Intent 过滤器的动作为"edu.xzceu.BroadcastReceiverDemo"，这与 BroadcastReceiverDemo.java 文件中 Intent 的动作相一致，表明这个 BroadcastReceiver 可以接收动作为"edu.xzceu.BroadcastReceiverDemo"的广播消息。MyBroadcastReceiver.java 文件创建了一个自定义的 BroadcastReceiver，其核心代码如下。

```
1. public class MyBroadcastReceiver extends BroadcastReceiver {
2.     @Override
3.     public void onReceive(Context context, Intent intent) {
```

```
4.        String msg = intent.getStringExtra("message");
5.        Toast.makeText(context, msg, Toast.LENGTH_SHORT).show();
6.    }
7. }
```

第 1 行代码首先继承了 BroadcastReceiver 类。第 3 行代码重载了 onReveive()函数,当接收到 AndroidManifest.xml 文件定义的广播消息后,程序将自动调用 onReveive()函数。第 4 行代码通过调用 getStringExtra()函数,从 Intent 中获取标识为 message 的字符串数据,并使用 Toast 将信息显示在屏幕上。第 5 行代码的 Toast 是一个显示提示信息的类,调用 makeText()函数可将提示信息短时间的浮现在用户界面之上。makeText()函数的第 1 个参数是上下文信息,第 2 个参数是需要显示的提示信息,第 3 个参数是显示的时间(Toast. LENGTH_SHORT 表示短时间显示,Toast. LENGTH_LONG 表示长时间显示),最后调用 show()方法将提示信息实际显示在界面之上。

习题与思考题

1. 简述 Intent 的定义和用途。
2. 简述 Intent 过滤器的定义和功能。
3. 简述 Intent 解析的匹配规则。
4. 编程实现下述功能:主界面上有一个"登录"按钮,点击"登录"按钮后打开一个新的 Activity;新的 Activity 上面有输入用户名和密码的控件,在用户关闭这个 Activity 后,将用户输入的用户名和密码传递到主界面中。

第 7 章

Android 后台服务

☆ 7.1 Service 简介
☆ 7.2 本地服务
☆ 7.3 远程服务

本章学习目标：了解 Service 的原理和用途；掌握进程内服务的管理方法，掌握服务的隐式启动和显式启动方法，了解线程的启动、挂起和停止方法；了解跨线程的界面更新方法掌握跨进程服务的绑定和调用方法，了解 AIDL 语言的用途和语法。

7.1 Service 简介

Service 是 Android 系统的后台服务组件，适用于开发无界面、长时间运行的应用功能。

1. 特点

没有用户界面，比 Activity 的优先级高，不会轻易被 Android 系统终止，即使 Service 被系统终止，在系统资源恢复后 Service 也将自动恢复运行状态，用于进程间通信（Inter Process Communication, IPC），解决两个不同 Android 应用程序进程之间的调用和通讯问题。

2. Service 生命周期

Service 生命周期包括：全生命周期，活动生命周期。onCreate()事件回调函数：Service 的生命周期开始，完成 Service 的初始化工作。onStart()事件回调函数：活动生命周期开始，但没有与之对应的"停止"函数，因此可以近似认为活动生命周期也是以 onDestroy()标志结束。onDestroy()事件回调函数：Service 的生命周期结束，释放 Service 所有占用的资源。Service 周期如图 7-1 所示。

图 7-1 Service 生命周期

3. 启动方式

通过调用 Context.startService()启动 Service，通过调用 Context.stopService()或 Service.stopSefl()停止 Service。Service 是由其他的组件启动的，但停止过程可以通过自身或其他组件完成。

如果仅以启动方式使用的 Service，这个 Service 需要具备自管理的能力，且不需要通过函数调用向外部组件提供数据或功能。

4. 绑定方式

通过服务链接（Connection）或直接获取 Service 中状态和数据信息，服务链接能够获取 Service 的对象，因此绑定 Service 的组件可以调用 Service 中实现的函数。使用 Service 的组件通过 Context.bindService()建立服务链接，通过 Context.unbindService()停止服务链接。如果在绑定过程中 Service 没有启动，Context.bindService()会自动启动 Service。

同一个 Service 可以绑定多个服务链接，这样可以同时为多个不同的组件提供服务。

启动方式和绑定方式的结合，这两种使用方法并不是完全独立的，在某些情况下可以混合使用。以 MP3 播放器为例，在后台工作的 Service 通过 Context.startService()启动某个特定音乐播放，但在播放过程中如果用户需要暂停音乐播放，则需要通过 Context.bindService()获取服务链接和 Service 对象，进而通过调用 Service 的对象中的函数，暂停音乐播放过程，并保存相关信息。在这种情况下，如果调用 Context.stopService()并不能够停止 Service，需要在所有的服务链接关闭后，Service 才能够真正的停止。

7.2 本地服务

7.2.1 服务管理

1. 注册 Service

服务管理主要指服务的启动和停止,首先介绍实现 Service 的最小代码集:

```
1. import android.app.Service;
2. import android.content.Intent;
3. import android.os.IBinder;
4.
5. public class RandomService extends Service{
6.     @Override
7.     public IBinder onBind(Intent intent) {
8.         return null;
9.     }
10. }
```

第 1 行到第 3 行引入必要包。第 5 行声明了 RandomService 继承 android.app.Service 类。在第 7 行到第 9 行重载了 onBind()函数。

onBind()函数是在 Service 被绑定后调用的函数,能够返回 Service 的对象,在后面的内容中会详细介绍。Service 的最小代码集并不能完成任何实际的功能,需要重载 onCreate()、onStart()和 onDestroy(),才使 Service 具有实际意义。

Android 系统在创建 Service 时,会自动调用 onCreate() 完成必要的初始化工作,在 Service 没有必要再存在时,系统会自动调用 onDestroy(),释放所有占用的资源。通过 Context.startService(Intent)启动 Service 时,onStart()则会被系统调用,Intent 会传递给 Service 一些重要的参数,Service 会根据实际情况选择需要重载上面的三个函数。

```
1. public class RandomService extends Service{
2.     @Override
3.     public void onCreate() {
4.         super.onCreate();
5.     }
6.     @Override
7.     public void onStart(Intent intent, int startId) {
8.         super.onStart(intent, startId);
9.     }
10.     @Override
```

```
11.      public void onDestroy(){
12.          super.onDestroy();
13.      }
14. }
```

第 4 行、第 8 行和第 12 行的代码重载 onCreate()、onStart()和 onDestroy()三个函数时，必须要在代码中调用父函数。

在 AndroidManifest.xml 文件中注册 Service，否则，这个 Service 根本无法启动，AndroidManifest.xml 文件中注册 Service 的代码如下。

```
<service android:name=".RandomService"/>。
```

使用＜service＞标签声明服务，其中的 android:name 表示的是 Service 的类名称，一定要与用户建立的 Service 类名称一致。

2. 启动 Service

启动 Service 的方式分为显示启动和隐式启动。

显示启动：在 Intent 中指明 Service 所在的类，并调用 startService(Intent)函数启动 Service，示例代码如下。

```
1. final Intent serviceIntent = new Intent(this,RandomService.class);
2. startService(serviceIntent);
```

在 Intent 中指明启动的 Service 在 RandomSrevice.class 中。

隐式启动：在注册 Service 时，声明 Intent-filter 的 action 属性。

```
1. <service android:name=".RandomService">
2.     <intent-filter>
3.         <action android:name="edu.xzceu.RandomService" />
4.     </intent-filter>
5. </service>
```

设置 Intent 的 action 属性，可以在不声明 Service 所在类的情况下启动服务。

隐式启动的代码如下。

```
1. final Intent serviceIntent = new Intent();
2. serviceIntent.setAction("edu.xzceu.RandomService");
3. startService(serviceIntent);
```

若 Service 和调用服务的组件在同一个应用程序中，可以使用显式启动或隐式启动，显式启动更加易于使用，且代码简洁。若服务和调用服务的组件在不同的应用程序中，则只能使用隐式启动。

停止 Service，将启动 Service 的 Intent 传递给 stopService(Intent)函数即可，示例代码如下。

```
stopService(serviceIntent);
```

在调用 startService(Intent)函数首次启动 Service 后，系统会先后调用 onCreate()和 onStart()。再次调用 startService(Intent)函数，系统则仅调用 onStart()，而不再调用 onCreate()。在调用 stopService(Intent)函数停止 Service 时，系统会调用 onDestroy()。无论调用过多少次 startService(Intent)，在调用 stopService(Intent)函数时，系统仅调用 onDestroy()一次。

3. 开发实例

示例 SimpleRandomServiceDemo 以显式启动服务在应用程序中建立 Service。在工程中创建 RandomService 服务，该服务启动后会产生一个随机数，使用 Toast 显示在屏幕上。"启动 Service"按钮调用 startService(Intent)函数，启动 RandomService 服务。"停止 Service"按钮调用 stopService(Intent)函数，停止 RandomService 服务。如图 7-2 所示。

图 7-2　SimpleRandomServiceDemo 界面

RandomService.java 文件的代码如下。

```
1.   package edu.xzceu.SimpleRandomServiceDemo;
2.   import android.app.Service;
3.   import android.content.Intent;
4.   import android.os.IBinder;
5.   import android.widget.Toast;
6.
7.   public class RandomService extends Service{
8.       @Override
9.       public void onCreate() {
```

```
10.        super.onCreate();
11.        Toast.makeText(this,"(1)调用onCreate()",
12.        Toast.LENGTH_LONG).show();
13.    }
14.
15.    @Override
16.    public void onStart(Intent intent, int startId) {
17.        super.onStart(intent, startId);
18.        Toast.makeText(this,"(2)调用onStart()",
19.        Toast.LENGTH_SHORT).show();
20.        double randomDouble = Math.random();
21.        String msg = "随机数:"+ String.valueOf(randomDouble);
22.        Toast.makeText(this,msg, Toast.LENGTH_SHORT).show();
23.    }
24.
25.    @Override
26.    public void onDestroy() {
27.        super.onDestroy();
28.        Toast.makeText(this,"(3)调用onDestroy()",
29.                Toast.LENGTH_SHORT).show();
30.    }
31.
32.    @Override
33.    public IBinder onBind(Intent intent) {
34.        return null;
35.    }
36. }
```

在onStart()函数中添加生产随机数的代码,第23行生产一个介于0和1之间的随机数,第24行代码构造供Toast显示的消息。

AndroidManifest.xml文件的代码如下。

```
1. <?xml version="1.0" encoding="utf-8"?>
2. <manifest
   xmlns:android="http://schemas.android.com/apk/res/android"
3.      package="edu.xzceu.SimpleRandomServiceDemo"
4.      android:versionCode="1"
5.      android:versionName="1.0">
```

```
6.    <application android:icon="@drawable/icon"
            android:label="@string/app_name">
7.        <activity android:name=".SimpleRandomServiceDemo"
8.            android:label="@string/app_name">
9.            <intent-filter>
10.               <action
    android:name="android.intent.action.MAIN" />
11.               <category
    android:name="android.intent.category.LAUNCHER" />
12.           </intent-filter>
13.       </activity>
14.       <service android:name=".RandomService"/>
15.   </application>
16.   <uses-sdk android:minSdkVersion="3" />
17. </manifest>
```

在调用的 AndroidManifest.xml 文件中,在<application>标签下,包含一个<activity>标签和一个<service>标签,在<service>标签中,声明了 RandomService 所在的类。
SimpleRandomServiceDemo.java 文件的代码如下。

```
1.  package edu.xzceu.SimpleRandomServiceDemo;
2.  import android.app.Activity;
3.  import android.content.Intent;
4.  import android.os.Bundle;
5.  import android.view.View;
6.  import android.widget.Button;
7.  public class SimpleRandomServiceDemo extends Activity {
8.      @Override
9.      public void onCreate(Bundle savedInstanceState) {
10.         super.onCreate(savedInstanceState);
11.         setContentView(R.layout.main);
12.         Button startButton = (Button)findViewById(R.id.start);
13.         Button stopButton = (Button)findViewById(R.id.stop);
14.         final Intent serviceIntent = new Intent(this, RandomService.class);
15.         startButton.setOnClickListener(new Button.OnClickListener(){
16.             public void onClick(View view){
17.                 startService(serviceIntent);
18.             }
19.         });
```

```
20.        stopButton.setOnClickListener(new Button.OnClickListener(){
21.            public void onClick(View view){
22.                stopService(serviceIntent);
23.            }
24.        });
25.    }
26. }
```

第 20 行和第 25 行分别是启动和停止 Service 的代码。

7.2.2 使用线程

任何耗时的处理过程都会降低用户界面的响应速度,甚至导致用户界面失去响应。当用户界面失去响应超过 5 秒钟,Android 系统会允许用户强行关闭应用程序。如图 7-3 所示。

图 7-3 失去响应界面

较好的解决方法是将耗时的处理过程转移到子线程上,这样可以避免负责界面更新的主线程无法处理界面事件,从而避免用户界面长时间失去响应。

线程是独立的程序单元,多个线程可以并行工作。在各线程系统中,每个中央处理器单独运行一个线程,因此线程是并行工作的。在单处理器系统中,处理器会给每个线程一小段时间,在这个时间内线程是被执行的,然后处理器执行下一个线程,这样就产生了线程并行运行的假象。无论线程是否真的并行工作,在宏观上可以认为子线程是独立于主线程,且能与主线程并行工作的程序单元。

使用线程实现 Java 的 Runnable 接口,并重载 run()方法。在 run()中放置代码的主体部分。

```
1. private Runnable backgroudWork = new Runnable(){
2.     @Override
3.     public void run() {
4.         //过程代码
5.     }
6. };
```

1. 使用线程

创建 Thread 对象，并将上面实现的 Runnable 对象作为参数传递给 Thread 对象。

```
1. private Thread workThread;
2. workThread = new Thread(null,backgroudWork,"WorkThread");
```

Thread 的构造函数中，第 1 个参数用来表示线程组，第 2 个参数是需要执行的 Runnable 对象，第 3 个参数是线程的名称。

调用 start()方法启动线程。

```
workThread.start();
```

线程在 run()方法返回后，线程就自动终止了。不推荐使用调用 stop()方法在外部终止线程，最好的方法是通知线程自行终止。一般调用 interrupt()方法通告线程准备终止，线程会释放它正在使用的资源，在完成所有的清理工作后自行关闭。

```
workThread.interrupt();
```

interrupt()方法并不能直接终止线程，仅是改变了线程内部的一个布尔字段，run()方法能够检测到这个布尔字段，从而知道何时应该释放资源和终止线程。在 run()方法的代码，一般通过 Thread.interrupted()方法查询线程是否被中断。

下面的代码是以 1 秒为间隔循环检测断线程是否被中断。

```
1. public void run() {
2.    while(! Thread.interrupted()){
3.        //过程代码
4.        Thread.sleep(1000);
5.    }
6. }
```

第 4 行代码使线程休眠 1 000 毫秒。当线程在休眠过程中被中断，则会产生 InterruptedException，在中断的线程上调用 sleep()方法，同样会产生 InterruptedException。

2. Thread.interrupted 方法功能

Thread.interrupted()通过捕获 InterruptedException 判断线程是否应被中断，并且在捕获到 InterruptedException 后，安全终止线程。

```
1. public void run() {
2.    try {
3.        while(true){
4.            //过程代码
5.            Thread.sleep(1000);
6.        }
7.    } catch (InterruptedException e) {
```

```
 8.         e.printStackTrace();
 9.      }
10. }
```

3. 使用 Handler 更新用户界面

Handler 允许将 Runnable 对象发送到线程的消息队列中,每个 Handler 对象绑定到一个单独的线程和消息队列上。当用户建立一个新的 Handler 对象,通过 post()方法将 Runnable 对象从后台线程发送到 GUI 线程的消息队列中。当 Runnable 对象通过消息队列后,这个 Runnable 对象将被运行。

```
 1. private static Handler handler = new Handler();
 2.
 3. public static void UpdateGUI(double refreshDouble){
 4.   handler.post(RefreshLable);
 5. }
 6. private static Runnable RefreshLable = new Runnable(){
 7.   @Override
 8.   public void run() {
 9.     //过程代码
10.   }
11. };
```

第 1 行代码建立了一个静态的 Handler 对象,但这个对象是私有的,因此外部代码并不能直接调用这个 Handler 对象。第 3 行 UpdateGUI()是公有的界面更新函数,后台线程通过调用该函数,将后台产生的数据 refreshDouble 传递到 UpdateGUI()函数内部,然后并直接调用 post()方法,将第 6 行的创建的 Runnable 对象传递给界面线程(主线程)的消息队列中。第 7 行到第 10 行代码是 Runnable 对象中需要重载的 run()函数,一般将界面更新代码放置在 run()函数中。

示例 ThreadRandomServiceDemo 使用线程持续产生随机数。点击"启动 Service"后,将启动后台线程。点击"停止 Service"后,将关闭后台线程。后台线程每 1 秒钟产生一个 0 到 1 之间的随机数,并通过 Handler 将产生的随机数显示在用户界面上。如图 7-4 所示。

图 7-4 ThreadRandomServiceDemo 界面

在 ThreadRandomServiceDemo 示例中,RandomService.java 文件是描述 Service 的文件,

用来创建线程、产生随机数和调用界面更新函数。RandomService.java 文件的完整代码如下。

```
1.  package edu.xzceu.ThreadRandomServiceDemo;
2.  import android.app.Service;
3.  import android.content.Intent;
4.  import android.os.IBinder;
5.  import android.widget.Toast;
6.  public class RandomService extends Service{
7.      private Thread workThread;
8.
9.      @Override
10.     public void onCreate() {
11.         super.onCreate();
12.         Toast.makeText(this,"(1) 调用 onCreate()",
13.                 Toast.LENGTH_LONG).show();
14.         workThread = new Thread(null,backgroudWork,"WorkThread");
15.     }
16.
17.     @Override
18.     public void onStart(Intent intent, int startId) {
19.         super.onStart(intent, startId);
20.         Toast.makeText(this,"(2) 调用 onStart()",
21.                 Toast.LENGTH_SHORT).show();
22.         if (! workThread.isAlive()){
23.             workThread.start();
24.         }
25.     }
26.
27.     @Override
28.     public void onDestroy() {
29.         super.onDestroy();
30.         Toast.makeText(this,"(3) 调用 onDestroy()",
31.                 Toast.LENGTH_SHORT).show();
32.         workThread.interrupt();
33.     }
34.
35.     @Override
36.     public IBinder onBind(Intent intent) {
```

```
37.        return null；
38.     }
39.
40.     private Runnable backgroudWork = new Runnable(){
41.        @Override
42.        public void run() {
43.           try {
44.              while(！Thread.interrupted()){
45.                 double randomDouble = Math.random();
46.                 ThreadRandomServiceDemo.UpdateGUI(randomDouble);
47.                 Thread.sleep(1000);
48.              }
49.           } catch (InterruptedException e) {
50.              e.printStackTrace();
51.           }
52.        }
53.     };
54. }
```

ThreadRandomServiceDemo.java 文件是界面的 Activity 文件，封装 Handler 的界面更新函数就在这个文件中，ThreadRandomServiceDemo.java 文件的完整代码如下。

```
1.  package edu.xzceu.ThreadRandomServiceDemo；
2.  import android.app.Activity；
3.  import android.content.Intent；
4.  import android.os.Bundle；
5.  import android.os.Handler；
6.  import android.view.View；
7.  import android.widget.Button；
8.  import android.widget.TextView；
9.  public class ThreadRandomServiceDemo extends Activity {
10.
11.    private static Handler handler = new Handler();
12.    private static TextView labelView = null;
13.    private static double randomDouble ;
14.
15.    public static void UpdateGUI(double refreshDouble){
16.       randomDouble = refreshDouble;
17.     handler.post(RefreshLable);
```

```
18.     }
19.
20.     private static Runnable RefreshLable = new Runnable(){
21.         @Override
22.         public void run() {
23.             labelView.setText(String.valueOf(randomDouble));
24.         }
25.     };
26.
27.     @Override
28.     public void onCreate(Bundle savedInstanceState) {
29.         super.onCreate(savedInstanceState);
30.         setContentView(R.layout.main);
31.         labelView = (TextView)findViewById(R.id.label);
32.         Button startButton = (Button)findViewById(R.id.start);
33.         Button stopButton = (Button)findViewById(R.id.stop);
34.         final Intent serviceIntent = new Intent(this, RandomService.class);
35.
36.         startButton.setOnClickListener(new Button.OnClickListener(){
37.             public void onClick(View view){
38.                 startService(serviceIntent);
39.             }
40.         });
41.
42.         stopButton.setOnClickListener(new Button.OnClickListener(){
43.             public void onClick(View view){
44.                 stopService(serviceIntent);
45.             }
46.         });
47.     }
48. }
```

7.2.3 使用 Service

以绑定方式使用 Service,能够获取到 Service 对象,不仅能够正常启动 Service,而且能够调用正在运行中的 Service 实现的公有方法和属性。为了使 Service 支持绑定,需要在 Service 类中重载 onBind()方法,并在 onBind()方法中返回 Service 对象,示例代码如下。

```
1. public class MathService extends Service{
2.     private final IBinder mBinder = new LocalBinder();
3. 
4.     public class LocalBinder extends Binder{
5.         MathService getService() {
6.             return MathService.this;
7.         }
8.     }
9. 
10.    @Override
11.    public IBinder onBind(Intent intent) {
12.        return mBinder;
13.    }
14. }
```

当 Service 被绑定时,系统会调用 onBind()函数,通过 onBind()函数的返回值,将 Service 对象返回给调用者。

第 11 行代码中可以看出,onBind()函数的返回值必须符合 IBinder 接口,因此在代码的第 2 行声明一个接口变量 mBinder,mBinder 符合 onBind()函数返回值的要求,因此将 mBinder 传递给调用者。IBinder 是用于进程内部和进程间过程调用的轻量级接口,定义了与远程对象交互的抽象协议,使用时通过继承 Binder 的方法实现。第 4 行代码继承 Binder,LocalBinder 是继承 Binder 的一个内部类。第 5 行代码实现了 getService()函数,当调用者获取到 mBinder 后,通过调用 getService()即可获取到 Service 的对象。

调用者通过 bindService()函数绑定服务,并在第 1 个参数中将 Intent 传递给 bindService()函数,声明需要启动的 Service。第 3 个参数 Context.BIND_AUTO_CREATE 表明只要绑定存在,就自动建立 Service;同时也告知 Android 系统,这个 Service 的重要程度与调用者相同,除非考虑终止调用者,否则不要关闭这个 Service。

```
1. final Intent serviceIntent = new Intent(this,MathService.class);
2. bindService(serviceIntent,mConnection,Context.BIND_AUTO_CREATE);
```

bindService()函数的第 2 个参数是 ServiceConnnection。当绑定成功后,系统将调用 ServiceConnnection 的 onServiceConnected()方法,而当绑定意外断开后,系统将调用 ServiceConnnection 中的 onServiceDisconnected 方法。

由上可知,以绑定方式使用 Service,调用者需要声明一个 ServiceConnnection,并重载内部的 onServiceConnected()方法和 onServiceDisconnected 方法。

```
1. private ServiceConnection mConnection = new ServiceConnection() {
2.     @Override
```

```
3.    public void onServiceConnected(ComponentName name, IBinder service)
      {
4.        mathService =((MathService.LocalBinder)service).getService();
5.    }
6.    @Override
7.    public void onServiceDisconnected(ComponentName name) {
8.        mathService = null;
9.    }
10. };
```

在第 4 行代码中,绑定成功后通过 getService()获取 Service 对象,这样便可以调用 Service 中的方法和属性。第 8 行代码将 Service 对象设置为 null,表示绑定意外失效,Service 实例不再可用。

取消绑定仅需要使用 unbindService()方法,并将 ServiceConnnection 传递给 unbindService()方法。

需注意的是,unbindService()方法成功后,系统并不会调用 onServiceDisconnected(),因为 onServiceDisconnected()仅在意外断开绑定时才被调用。

```
unbindService(mConnection);
```

通过 bindService()函数绑定 Servcie 时,onCreate()函数和 onBinde()函数将先后被调用。通过 unbindService()函数取消绑定 Servcie 时,onUnbind()函数将被调用。如果 onUnbind()函数的返回 true,则表示在调用者绑定新服务时,onRebind()函数将被调用。

绑定方式的函数调用顺序如图 7-5 所示。

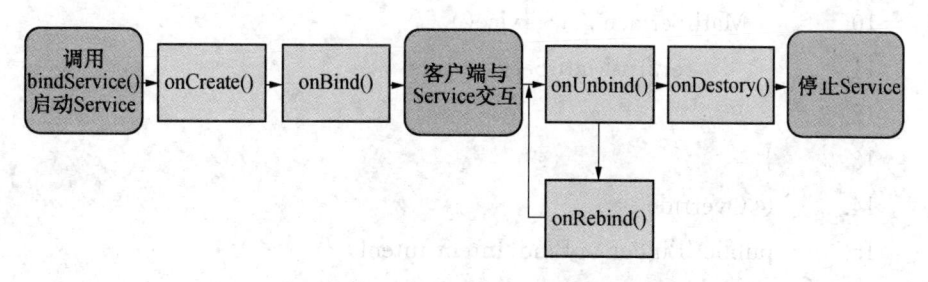

图 7-5 绑定方式的函数调用顺序

下面示例使用 SimpleMathServiceDemo 绑定方式使用 Service。

创建 MathService 服务,用来完成简单的数学运算,但足以说明如何使用绑定方式调用 Service 实例中的公有方法。如图 7-6 所示。

在服务绑定后,用户可以点击"加法运算",将两个随机产生的数值传递给 MathService 服务,并从 MathService 对象中获取到加法运算的结果,然后显示在屏幕的上方。"取消绑定"按钮可以解除与 MathService 的绑定关系,取消绑定后,无法通过"加法运算"按钮获取加法运算结果。如图 7-6 所示。

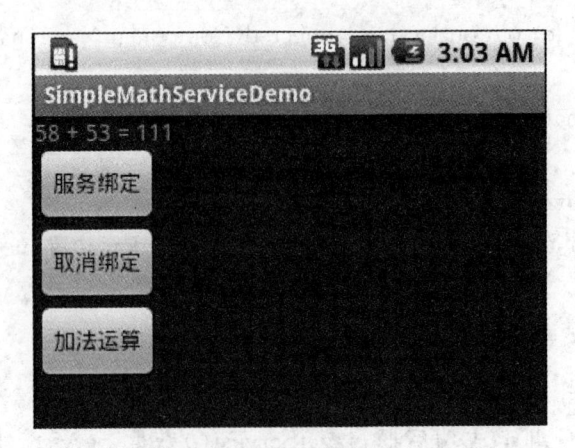

图 7-6 SimpleMathServiceDemo 示例

在 SimpleMathServiceDemo 示例中，MathService.java 文件是描述 Service 的文件，MathService.java 文件的完整代码如下。

```
1. package edu.xzceu.SimpleMathServiceDemo;
2. import android.app.Service;
3. import android.content.Intent;
4. import android.os.Binder;
5. import android.os.IBinder;
6. import android.widget.Toast;
7. public class MathService extends Service{
8.     private final IBinder mBinder = new LocalBinder();
9.     public class LocalBinder extends Binder{
10.        MathService getService() {
11.            return MathService.this;
12.        }
13.    }
14.    @Override
15.    public IBinder onBind(Intent intent) {
16.        Toast.makeText(this,"本地绑定:MathService",
17.            Toast.LENGTH_SHORT).show();
18.        return mBinder;
19.    }
20.    @Override
21.    public boolean onUnbind(Intent intent){
22.        Toast.makeText(this,"取消本地绑定:MathService",
```

```
23.            Toast.LENGTH_SHORT).show();
24.        return false;
25.     }
26.     public long Add(long a,long b){
27.        return a+b;
28.     }
29. }
```

SimpleMathServiceDemo.java 文件是界面的 Activity 文件,绑定和取消绑定服务的代码在这个文件中,SimpleMathServiceDemo.java 文件的完整代码如下。

```
1.  package edu.xzceu.SimpleMathServiceDemo;
2.  import android.app.Activity;
3.  import android.content.ComponentName;
4.  import android.content.Context;
5.  import android.content.Intent;
6.  import android.content.ServiceConnection;
7.  import android.os.Bundle;
8.  import android.os.IBinder;
9.  import android.view.View;
10. import android.widget.Button;
11. import android.widget.TextView;
12. public class SimpleMathServiceDemo extends Activity {
13.     private MathService mathService;
14.     private boolean isBound = false;
15.     TextView labelView;
16.     @Override
17.     public void onCreate(Bundle savedInstanceState) {
18.        super.onCreate(savedInstanceState);
19.        setContentView(R.layout.main);
20.        labelView = (TextView)findViewById(R.id.label);
21.        Button bindButton = (Button)findViewById(R.id.bind);
22.        Button unbindButton = (Button)findViewById(R.id.unbind);
23.        Button computButton = (Button)findViewById(R.id.compute);
24.        bindButton.setOnClickListener(new View.OnClickListener(){
25.           @Override
```

```java
26.     public void onClick(View v) {
27.        if(! isBound){
28.           final Intent serviceIntent = newIntent
    (SimpleMathServiceDemo.this,MathService.class);
29.           bindService(serviceIntent,mConnection,Context.
    BIND_AUTO_CREATE);
30.           isBound = true;
31.        }
32.     }
33.  });
34.
35.  unbindButton.setOnClickListener(new View.OnClickListener(){
36.     @Override
37.     public void onClick(View v) {
38.        if(isBound){
39.           isBound = false;
40.           unbindService(mConnection);
41.           mathService = null;
43.        }
44.     }
44.  });
45.
46.  computButton.setOnClickListener(new View.OnClickListener(){
47.     @Override
48.     public void onClick(View v) {
49.        if (mathService == null){
50.           labelView.setText("未绑定服务");
51.           return;
52.        }
53.        long a = Math.round(Math.random() * 100);
54.        long b = Math.round(Math.random() * 100);
55.        long result = mathService.Add(a, b);
56.        String msg = String.valueOf(a)+" + "+String.valueOf(b)+
57.           " = "+String.valueOf(result);
58.        labelView.setText(msg);
59.     }
```

```
60.        });
61.    }
62.
63.    private ServiceConnection mConnection = new ServiceConnection() {
64.        @Override
65.        public void onServiceConnected(ComponentName name,
            IBinder service) {
66.            mathService = ((MathService.LocalBinder)service).getService();
67.        }
68.
69.        @Override
70.        public void onServiceDisconnected(ComponentName name) {
71.            mathService = null;
72.        }
73.    };
74. }
```

7.3 远程服务

7.3.1 进程间通信

在 Android 系统中,每个应用程序在各自的进程中运行,而且出于安全原因的考虑,这些进程之间彼此是隔离的,进程之间传递数据和对象,需要使用 Android 支持的进程间通信(Inter-Process Communication,IPC)机制。进程间通信采用 Intent 和跨进程服务的方式实现 IPC,使应用程序具有更好的独立性和健壮性。

IPC 机制包含:(1) Intent:承载数据,是一种简单、高效、且易于使用的 IPC 机制。(2) 远程服务:服务和调用者在不同的两个进程中,调用过程需要跨越进程才能实现。

实现远程服务的步骤:使用 AIDL 语言定义跨进程服务的接口;根据 AIDL 语言定义的接口,在具体的 Service 类中实现接口中定义的方法和属性;在需要调用跨进程服务的组件中,通过相同的 AIDL 接口文件,调用跨进程服务。

7.3.2 服务创建与调用

AIDL(Android Interface Definition Language)是 Android 系统自定义的接口描述语言,可以简化进程间数据格式转换和数据交换的代码,通过定义 Service 内部的公共方法,允许调用者和 Service 在不同进程间相互传递数据。AIDL 的 IPC 机制与 COM 和 Corba 非常相似，都是基于接口的轻量级进程通信机制。AIDL 语言的语法与 Java 语言的接口定义非常相似，唯一不同之处是：AIDL 允许定义函数参数的传递方向。

AIDL 支持三种方向：in、out 和 inout。标识为 in 的参数将从调用者传递到跨进程服务中，标识为 out 的参数将从跨进程服务传递到调用者中，标识为 inout 的参数将先从调用者传递到跨进程服务中，再从跨进程服务返回给调用者。在不标识参数的传递方向时，缺省认定所有函数的传递方向为 in。出于性能方面的考虑，不要在参数中标识不需要的传递方向。

远程访问的创建和调用需要使用 AIDL 语言，一般分为以下几个过程：使用 AIDL 语言定义跨进程服务的接口，通过继承 Service 类实现跨进程服务，绑定和使用跨进程服务。

下面以 RemoteMathServiceDemo 示例为参考，说明如何创建跨进程服务。在这个示例中，仅定义了 MathService 服务，可以为远程调用者提供加法服务。

使用 AIDL 语言定义跨进程服务的接口。首先使用 AIDL 语言定义的 MathService 的服务接口，文件名为 IMathService.aidl。

IMathService 接口仅包含一个 add() 方法，传入的参数是两个长型整数，返回值也是长型整数。

```
1. package com.dh.service;
2. interface IMathService {
3. long Add(long a, long b);
4. }
```

1. 使用 AIDL 语言定义跨进程服务的接口

如果使用 Eclipse 编辑 IMathService.aidl 文件，当用户保存文件后，ADT 会自动在 /gen 目录下生成 IMathService.java 文件，图 7-7 是 IMathService.java 文件结构。

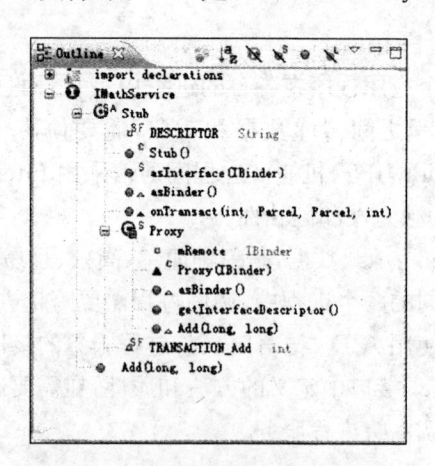

图 7-7　IMathService.java 文件结构

使用 AIDL 语言定义跨进程服务的接口。IMathService.java 文件根据 IMathService.aidl 的定义，生成了一个内部静态抽象类 Stub，Stub 继承了 Binder 类，并实现 ImathService 接口。在 Stub 类中，还包含一个重要的静态类 Proxy。如果认为 Stub 类实现进程内服务调用，那么

Proxy 类则是用来实现跨进程服务调用的,将 Proxy 作为 Stub 的内部类完全是出于使用方便的目的。Stub 类和 Proxy 类关系图如图 7-8 所示。

```
                    AIDL文件
                  IMath Service.aidl
                         │
                       AIDL
                       工具
                         ▼
        ┌─────────────────────────────────────┐
        │   生成Java接口文件                    │
        │   IMathService.java                 │
        │   ┌─────────────────────────────┐   │
        │   │  生成内部静态抽象 Stub类      │   │
        │   │  IMathService.Stub          │   │
        │   │  ┌───────────────────────┐  │   │
        │   │  │ 生成内部静态抽象 Proxy类│  │   │
        │   │  │ IMathService.Stub.Proxy│  │   │
        │   │  └───────────────────────┘  │   │
        │   └─────────────────────────────┘   │
        └─────────────────────────────────────┘
           ↓                              ↓
    ╭──────────────╮              ╭──────────────╮
    │  本地服务对象  │              │  远程服务对象  │
    │Stub.asInterface()│          │使用asInterface()获取│
    │用来返回远程服务  │            │远程Proxy对象的引用│
    │对象(Proxy)     │              │              │
    │ onTransact()  │              │  Transact()  │
    ╰──────────────╯              ╰──────────────╯

    ┌────────────────────────────┐  ┌────────────────────────────┐
    │     IMathService.Stub      │  │   IMathService.Stub.Proxy  │
    ├────────────────────────────┤  ├────────────────────────────┤
    │IMathService asInterface    │  │IBinder asBinder()          │
    │          (IBinder obj)     │  │String getInterfaceDescriptor()│
    │IBinder asBinder()          │  │long Add(long a,long b)     │
    │boolean onTransact(int code,│  │                            │
    │  Parcel data,              │  │                            │
    │  Parcel reply,int flags)   │  │                            │
    └────────────────────────────┘  └────────────────────────────┘
```

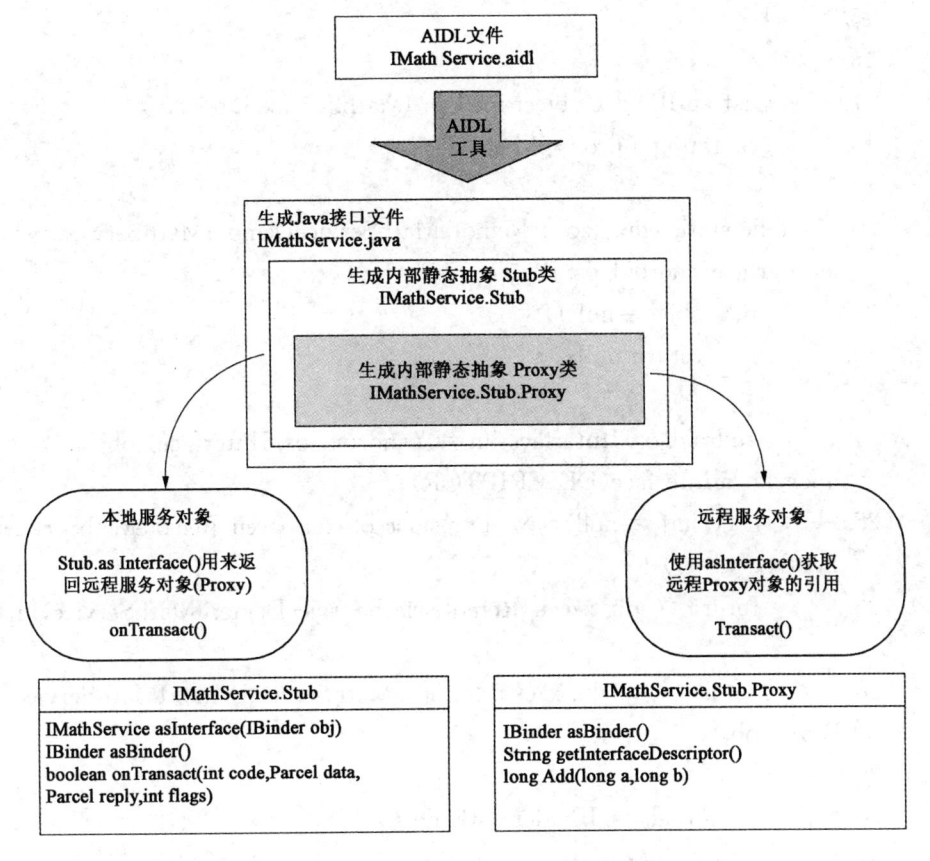

图 7-8 Stub 类和 Proxy 类关系图

IMathService.java 的完整代码如下。

```
1.  package edu.xzceu.RemoteMathServiceDemo;
2.  import java.lang.String;
3.  import android.os.RemoteException;
4.  import android.os.IBinder;
5.  import android.os.IInterface;
6.  import android.os.Binder;
7.  import android.os.Parcel;
8.  public interface IMathService extends android.os.IInterface{
9.  /** Local-side IPC implementation stub class. */
10. public static abstract class Stub extends android.os.Binder implements edu.
    xzceu.RemoteMathServiceDemo.IMathService{
11.   private static final java.lang.String DESCRIPTOR = "edu.xzceu.
    RemoteMathServiceDemo.IMathService";
12. /** Construct the stub at attach it to the interface. */
```

```
13.    public Stub(){
14.        this.attachInterface(this, DESCRIPTOR);
15.    }
16.    /**
17.     * Cast an IBinder object into an IMathService interface,
18.     * generating a proxy if needed.
19.     */
20.    public static edu.xzceu.RemoteMathServiceDemo.IMathService asInterface(android.os.IBinder obj){
21.        if ((obj==null)) {
22.            return null;
23.        }
24.        android.os.IInterface iin = (android.os.IInterface) obj.queryLocalInterface(DESCRIPTOR);
25.        if (((iin!=null)&&(iin instanceof edu.xzceu.RemoteMathServiceDemo.IMathService))) {
26.            return ((edu.xzceu.RemoteMathServiceDemo.IMathService)iin);
27.        }
28.        return new edu.xzceu.RemoteMathServiceDemo.IMathService.Stub.Proxy(obj);
29.    }
30.    public android.os.IBinder asBinder(){
31.        return this;
32.    }
33.    public boolean onTransact(int code, android.os.Parcel data, android.os.Parcel reply, int flags) throws android.os.RemoteException{
34.        switch (code){
35.        case INTERFACE_TRANSACTION:
36.        {
37.            reply.writeString(DESCRIPTOR);
38.            return true;
39.        }
40.        case TRANSACTION_Add:
41.        {
42.            data.enforceInterface(DESCRIPTOR);
43.            long _arg0;
44.            _arg0 = data.readLong();
45.            long _arg1;
```

```
46.        _arg1 = data.readLong();
47.        long _result = this.Add(_arg0, _arg1);
48.        reply.writeNoException();
49.        reply.writeLong(_result);
50.        return true;
51.      }
52. }
53. return super.onTransact(code, data, reply, flags);
54. }
55.   private static class Proxy implements edu.xzceu.RemoteMathServiceDemo.IMathService{
56.   private android.os.IBinder mRemote;
57.   Proxy(android.os.IBinder remote){
58.       mRemote = remote;
59.   }
60.   public android.os.IBinder asBinder(){
61.       return mRemote;
62.   }
63.   public java.lang.String getInterfaceDescriptor(){
64.       return DESCRIPTOR;
65.   }
66.   public long Add(long a, long b) throws android.os.RemoteException{
67.   android.os.Parcel _data = android.os.Parcel.obtain();
68.         android.os.Parcel _reply = android.os.Parcel.obtain();
69.       long _result;
70.       try {
71.         _data.writeInterfaceToken(DESCRIPTOR);
72.         _data.writeLong(a);
73.         _data.writeLong(b);
74.         mRemote.transact(Stub.TRANSACTION_Add, _data, _reply, 0);
75.         _reply.readException();
76.         _result = _reply.readLong();
77.       }
78.       finally {
79.         _reply.recycle();
80.         _data.recycle();
81.       }
82.   return _result;
83.   }
```

```
84.    }
85.    static final int TRANSACTION_Add = (IBinder. FIRST_CALL_
       TRANSACTION + 0);
86.  }
87.  public long Add(long a, long b) throws android. os. RemoteException;
88. }
```

第 8 行代码是 IMathService 继承了 android. os. IInterface,这是所有使用 AIDL 建立的接口都必须继承的基类接口,这个基类接口中定义了 asBinder()方法,用来获取 Binder 对象。在代码的第 30 行到第 32 行,实现了 android. os. IInterface 接口所定义的 asBinder()方法。

在 IMathService 中,绝大多数的代码是用来实现 Stub 这个抽象类的。每个远程接口都包括 Stub 类,因为是内部类,所以并不会产生命名的冲突。asInterface(IBinder)是 Stub 内部的跨进程服务接口,调用者可以通过该方法获取到跨进程服务的对象。

仔细观察 asInterface(IBinder)实现方法,首先判断 IBinder 对象 obj 是否为 null(第 21 行),如果是则立即返回。第 24 行代码是使用 DESCRIPTOR 构造 android. os. IInterface 对象,并判断 android. os. IInterface 对象是否为进程内服务,如果是进程内服务,则无需进程间通信,返回 android. os. IInterface 对象(第 26 行);如果不是进程内服务,则构造并返回 Proxy 对象(第 28 行)。第 66 行代码是 Proxy 内部包含与 IMathService. aidl 相同签名的函数。第 71~76 行代码是在该函数中以一定的顺序将所有参数写入 Parcel 对象,以供 Stub 内部的 onTransact()方法能够正确获取到参数。

当数据以 Parcel 对象的形式传递到跨进程服务的内部时,onTransact()方法(第 33 行)将从 Parcel 对象中逐一的读取每个参数,然后调用 Service 内部制定的方法,并再将结果写入另一个 Parcel 对象,准备将这个 Parcel 对象返回给远程的调用者。Parcel 是 Android 系统中应用程序进程间数据传递的容器,能够在两个进程中完成数据的打包和拆包的工作。但 Parcel 不同于通用意义上的序列化,Parcel 的设计目的是用于高性能 IPC 传输,因此不能够将 Parcel 对象保存在任何持久存储设备上。

2. 通过继承 Service 类实现跨进程服务

实现跨进程服务需要建立一个继承 android. app. Service 的类,并在该类中通过 onBind()方法返回 IBinder 对象,调用者使用返回的 IBinder 对象就可以访问跨进程服务。IBinder 对象的建立通过使用 IMathService. java 内部的 Stub 类实现,并逐一实现在 IMathService. aidl 接口文件定义的函数。

在 RemoteMathServiceDemo 示例中,跨进程服务的实现类是 MathService. java。下面是 MathService. java 的完整代码。

```
1.  package edu. xzceu. RemoteMathServiceDemo;
2.
3.  import android. app. Service;
4.  import android. content. Intent;
5.  import android. os. IBinder;
6.  import android. widget. Toast;
```

```
7.
8.    public class MathService extends Service{
9.        private final IMathService.Stub mBinder = new IMathService.Stub() {
10.           public long Add(long a, long b) {
11.               return a + b;
12.           }
13.       };
14.       @Override
15.       public IBinder onBind(Intent intent) {
16.           Toast.makeText(this,"远程绑定:MathService",
17.                   Toast.LENGTH_SHORT).show();
18.           return mBinder;
19.       }
20.       @Override
21.       public boolean   onUnbind  (Intent intent){
22.           Toast.makeText(this,"取消远程绑定:MathService",
23.                   Toast.LENGTH_SHORT).show();
24.           return false;
25.       }
26.   }
```

第 8 行代码表明 MathService 继承于 android.app.Service。第 9 行代码建立 IMathService.Stub 的对象 mBinder。第 10 行代码实现了 AIDL 文件定义的跨进程服务接口。第 18 行代码在 onBind()方法中,将 mBinder 返回给远程调用者。第 16 行和第 22 行代码分别是在绑定和取消绑定时,为用户产生的提示信息。

RemoteMathServiceDemo 示例的文件结构如图 7-9 所示。

示例中只有跨进程服务的类文件 MathService.java 和接口文件 IMathService.aidl,没有任何用于启动时显示用户界面的 Activity 文件。

在调试 RemoteMathServiceDemo 示例时,模拟器的屏幕上不会出现用户界面,但在控制台会有"没有找到用于启动的 Activity,仅将应用程序同步到设备上"的提示信息,这些信息表明 apk 文件已经上传到模拟器中。提示信息如图 7-10 所示。

图 7-9　RemoteMathServiceDemo 示例的文件结构

```
Android Launch!
adb is running normally.
No Launcher activity found!
The launch will only sync the application package on the device!
Performing sync
Automatic Target Mode: using existing emulator 'emulator-5554' running compatible
Uploading RemoteMathServiceDemo.apk onto device 'emulator-5554'
Installing RemoteMathServiceDemo.apk...
Application already exists. Attempting to re-install instead...
Success!
\RemoteMathServiceDemo\bin\RemoteMathServiceDemo.apk installed on device
Done!
```

图 7-10　提示信息

使用 File Explorer 查看模拟器的文件系统，可以进一步确认编译好的 apk 文件是否正确上传到模拟器中。如果能在/data/app/下找到 edu.xzceu.RemoteMathServiceDemo.apk 文件，说明提供跨进程服务的 apk 文件已经正确上传，如果 RemoteMathServiceDemo 示例无法在 Android 模拟器的程序启动栏中找到，只能够通过其他应用程序调用该示例中的跨进程服务。

如下图 7-11 表示 edu.xzceu.RemoteMathServiceDemo.apk 文件的保存位置。

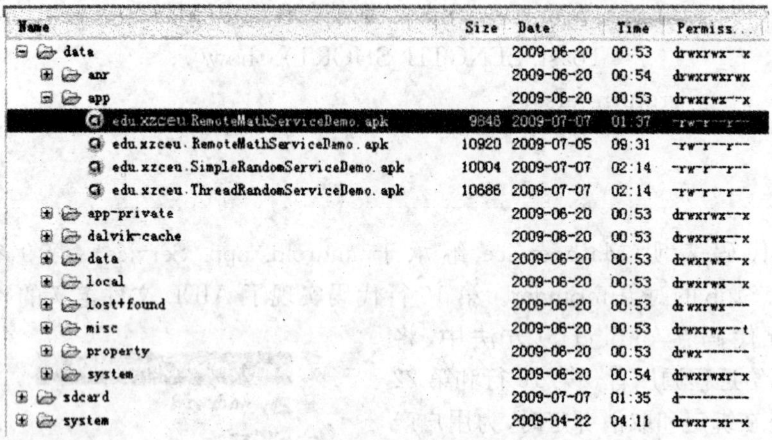

图 7-11　edu.xzceu.RemoteMathServiceDemo.apk 文件的保存位置

RemoteMathServiceDemo 是我们接触到的第一个没有 Activity 的示例，在 AndroidManifest.xml 文件中，在 <application> 标签下只有一个 <service> 标签。AndroidManifest.xml 文件的完整代码如下。

```
1.    <? xml version="1.0" encoding="utf-8"? >
2.    <manifest
      xmlns:android="http://schemas.android.com/apk/res/android"
3.    package="edu.xzceu.RemoteMathServiceDemo"
4.    android:versionCode="1"
5.    android:versionName="1.0">
6.    <application android:icon="@drawable/icon" android:label= "@string/app_name">
```

```
7.     <service android:name=".MathService"
8.         android:process=":remote">
9.         <intent-filter>
10.            <action android:name= "edu.xzceu.RemoteMathServiceDemo.
               MathService" />
11.        </intent-filter>
12.     </service>
13. </application>
14. <uses-sdk android:minSdkVersion="3" />
15. </manifest>
```

注意第 10 行代码，edu.xzceu.RemoteMathServiceDemo.MathService 是远程调用 MathService 的标识，在调用者段使用 Intent.setAction()函数将标识加入 Intent 中，然后隐式启动或绑定服务。

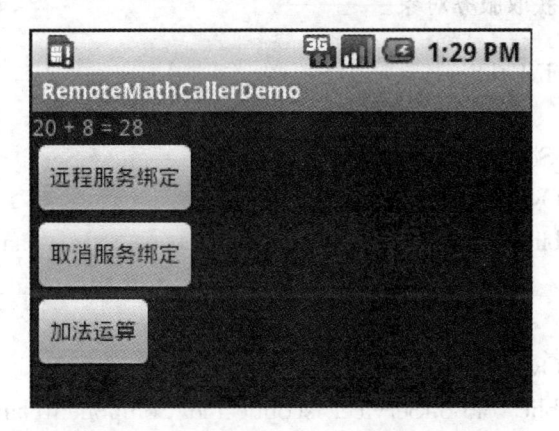

下图 7 - 12 是 RemoteMathCallerDemo 的界面，用户可以绑定跨进程服务，也可以取消服务绑定。在绑定跨进程服务后，可以调用 RemoteMathServiceDemo 中的 MathService 服务进行加法运算，运算的输入由 RemoteMathCallerDemo 随机产生，运算的输入和结果显示在屏幕的上方。

图 7 - 12　RemoteMathCallerDemo 的界面

3. 绑定和使用跨进程服务

应用程序在调用跨进程服务时，应用程序与跨进程服务应具有相同的 Proxy 类和签名函数，这样才能够使数据在调用者处打包后，可以在远程访问端正确拆包，反之亦然。从实践角度来讲，调用者需要使用与跨进程服务端相同的 AIDL 文件。在 RemoteMathCallerDemo 示例，在 edu.xzceu.RemoteMathServiceDemo 包下，引如与 RemoteMathServiceDemo 相同的 AIDL 文件 IMathService.aidl，同时在/gen 目录下会自动产生相同的 IMathService.java 文件。

下图 7 - 13 是 RemoteMathServiceDemo 的文件结构。

```
RemoteMathCallerDemo
  Android
  src
    edu.xzceu.RemoteMathCallerDemo
      RemoteMathCallerDemo.java
    edu.xzceu.RemoteMathServiceDemo
      IMathService.aidl
  gen [Generated Java Files]
    edu.xzceu.RemoteMathCallerDemo
      R.java
    edu.xzceu.RemoteMathServiceDemo
      IMathService.java
  assets
  res
  AndroidManifest.xml
  default.properties
  AndroidManifest.xml
  default.properties
```

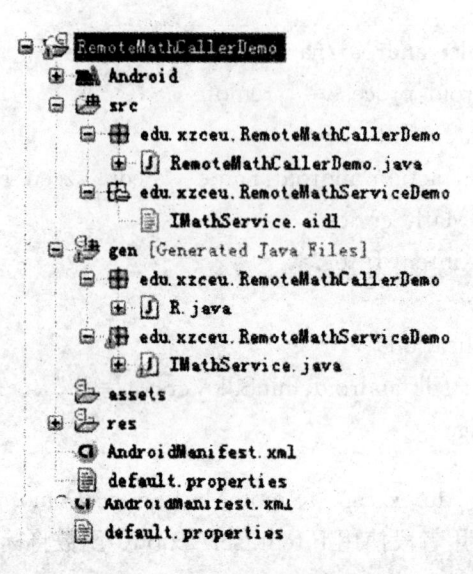

图 7-13 RemoteMathServiceDemo 的文件结构

RemoteMathCallerDemo.java 是 Activity 的文件,跨进程服务的绑定和使用方法与 7.2.3 节的进程内服务绑定示例 SimpleMathServiceDemo 相似,不同之处主要包括以下两个方面:第 1 行代码使用 IMathService 声明跨进程服务对象,第 6 行代码通过 IMathService.Stub 的 asInterface()方法实现获取服务对象。

```
1.  private IMathService mathService;
2.
3.  private ServiceConnection mConnection = new ServiceConnection() {
4.      @Override
5.      public void onServiceConnected(ComponentName name, IBinder service) {
6.          mathService = IMathService.Stub.asInterface(service);
7.      }
8.      @Override
9.      public void onServiceDisconnected(ComponentName name){
10.         mathService = null;
11.     }
12. };
```

绑定服务时,首先通过 setAction()方法声明服务标识,然后调用 bindService()绑定服务。服务标识必须与跨进程服务在 AndroidManifest.xml 文件中声明的服务标识完全相同。因此本示例的服务标识为 edu.xzceu.RemoteMathServiceDemo.MathService,与跨进程服务示例 RemoteMathServiceDemo 在 AndroidManifest.xml 文件声明的服务标识一致。

```
1.  final Intent serviceIntent = new Intent();
2.  serviceIntent.setAction("edu.xzceu.RemoteMathServiceDemo.MathService");
3.  bindService(serviceIntent,mConnection,Context.BIND_AUTO_CREATE);
```

下面是 RemoteMathCallerDemo.java 文件的完整代码。

```java
1.   package edu.xzceu.RemoteMathCallerDemo;
2.   import edu.xzceu.RemoteMathServiceDemo.IMathService;
3.   import android.app.Activity;
4.   import android.content.ComponentName;
5.   import android.content.Context;
6.   import android.content.Intent;
7.   import android.content.ServiceConnection;
8.   import android.os.Bundle;
9.   import android.os.IBinder;
10.  import android.os.RemoteException;
11.  import android.view.View;
12.  import android.widget.Button;
13.  import android.widget.TextView;
14.
15.  public class RemoteMathCallerDemo extends Activity {
16.      private IMathService mathService;
17.
18.      private ServiceConnection mConnection = new ServiceConnection() {
19.        @Override
20.        public void onServiceConnected(ComponentName name, IBinder service) {
21.            mathService = IMathService.Stub.asInterface(service);
22.        }
23.        @Override
24.        public void onServiceDisconnected(ComponentName name) {
25.            mathService = null;
26.        }
27.      };
28.
29.      private boolean isBound = false;
30.      TextView labelView;
31.      @Override
32.      public void onCreate(Bundle savedInstanceState) {
33.          super.onCreate(savedInstanceState);
34.          setContentView(R.layout.main);
35.
36.          labelView = (TextView)findViewById(R.id.label);
37.          Button bindButton = (Button)findViewById(R.id.bind);
```

```
38.        Button unbindButton = (Button)findViewById(R.id.unbind);
39.        Button computButton = (Button)findViewById(R.id.compute_
   add);
40.
41.        bindButton.setOnClickListener(new View.OnClickListener(){
42.            @Override
43.            public void onClick(View v) {
44.                if(! isBound){
45.                    final Intent serviceIntent = new Intent();
46.                    serviceIntent.setAction("edu.xzceu.RemoteMathServiceDemo.
   MathService");
47.                    bindService(serviceIntent, mConnection, Context.BIND_
   AUTO_CREATE);
48.                    isBound = true;
49.                }
50.            }
51.        });
52.
53.        unbindButton.setOnClickListener(new View.OnClickListener(){
54.            @Override
55.            public void onClick(View v) {
56.                if(isBound){
57.                    isBound = false;
58.                    unbindService(mConnection);
59.                    mathService = null;
60.                }
61.            }
62.        });
63.
64.        computButton.setOnClickListener(new View.OnClickListener(){
65.            @Override
66.            public void onClick(View v) {
67.                if (mathService == null){
68.                    labelView.setText("未绑定跨进程服务");
69.                    return;
70.                }
71.                long a = Math.round(Math.random() * 100);
72.                long b = Math.round(Math.random() * 100);
73.                long result = 0;
```

```
74.            try {
75.                result = mathService.Add(a, b);
76.            } catch (RemoteException e) {
77.                e.printStackTrace();
78.            }
79.            String msg = String.valueOf(a)+" + "+String.valueOf(b)+
80.                " = "+String.valueOf(result);
81.            labelView.setText(msg);
82.        }
83.    });
84. }
85. }
```

7.3.3 进程间传递

在 Android 系统中,进程间传递的数据包括:Java 语言支持的基本数据类型和用户自定义的数据类型。所有数据都必须是"可打包"的,才能够穿越进程边界。Java 语言的基本数据类型的打包过程是自动完成。对于自定义的数据类型,则需要实现 Parcelable 接口,使自定义的数据类型能够转换为系统级原语保存在 Parcel 对象中,穿越进程边界后可再转换为初始格式。AIDL 支持的数据类型表如表 7-1 所示。

表 7-1 数据类型表

类 型	说 明	需要引入
Java 语言的基本类型	包括 boolean、byte、short、int、float 和 double 等	否
String	java.lang.String	否
CharSequence	java.lang.CharSequence	否
List	其中所有的元素都必须是 AIDL 支持的数据类型	否
Map	其中所有的键和元素都必须是 AIDL 支持的数据类型	否
其他 AIDL 接口	任何其他使用 AIDL 语言生成的接口类型	是
Parcelable 对象	实现 Parcelable 接口的对象	是

下面以 ParcelMathServiceDemo 示例为参考,说明如何在跨进程服务中使用自定义数据类型。

这个示例是 RemoteMathServiceDemo 示例的延续,定义 MathService 服务,为远程调用者提供加法服务,没有启动界面,因此在模拟器的调试过程与 RemoteMathServiceDemo 示例相同。不同之处在于 MathService 服务增加了"全运算"功能,在接收到输入参数后,将向调用者返回一个包含"加、减、乘、除"全部运算结果的对象。这个对象是一个自定义的类,为了能够使其他 AIDL 文件可使用这个自定义类,需要使用 AIDL 语言声明这个类。
首先建立 AllResult. aidl 文件,声明 AllResult 类。

```
1.    package edu. xzceu. ParcelMathServiceDemo;
2.    parcelable AllResult;
```

在第 2 行代码中使用 parcelable 声明自定义类,这样其他的 AIDL 文件就可以使用这个自定义的类。

图 7-14 是 ParcelMathServiceDemo 的文件结构。

```
ParcelMathServiceDemo
├── Android
├── src
│   └── edu.xzceu.ParcelMathServiceDemo
│       ├── AllResult.java
│       ├── MathService.java
│       ├── AllResult.aidl
│       └── IMathService.aidl
├── gen [Generated Java Files]
│   └── edu.xzceu.ParcelMathServiceDemo
│       ├── IMathService.java
│       └── R.java
├── assets
├── res
├── AndroidManifest.xml
└── default.properties
```

图 7-14 ParcelMathServiceDemo 的文件结构

下面是 IMathService. aidl 文件的代码。

```
1.    package edu. xzceu. ParcelMathServiceDemo;
2.    import edu. xzceu. ParcelMathServiceDemo. AllResult;
3.
4.    interface IMathService {
5.        long Add(long a, long b);
6.        AllResult ComputeAll(long a, long b);
7.    }
```

第 2 行代码中引入了 edu. xzceu. ParcelMathServiceDemo. AllResult,才能够使用自定义数据结构 AllResult。第 6 行代码为全运算增加了新的函数 ComputeAll(),该函数的返回值就是在 AllResult. aidl 文件中定义的 AllResult。

构造 AllResult 类。AllResult 类除了基本的构造函数以外,还需要有以 Parcel 对象为输入的构造函数,并且需要重载打包函数 writeToParcel()。AllResult. java 文件的完整代码如下。

```java
1.  package edu.xzceu.ParcelMathServiceDemo;
2.  import android.os.Parcel;
3.  import android.os.Parcelable;
4.  public class AllResult implements Parcelable {
5.      public long AddResult;
6.      public long SubResult;
7.      public long MulResult;
8.      public double DivResult;
9.
10.     public AllResult(long addRusult, long subResult, long mulResult, double divResult){
11.         AddResult = addRusult;
12.         SubResult = subResult;
13.         MulResult = mulResult;
14.         DivResult = divResult;
15.     }
16.
17.     public AllResult(Parcel parcel) {
18.         AddResult = parcel.readLong();
19.         SubResult = parcel.readLong();
20.         MulResult = parcel.readLong();
21.         DivResult = parcel.readDouble();
22.     }
23.
24.     @Override
25.     public int describeContents() {
26.         return 0;
27.     }
28.
29.     @Override
30.     public void writeToParcel(Parcel dest, int flags) {
31.         dest.writeLong(AddResult);
32.         dest.writeLong(SubResult);
33.         dest.writeLong(MulResult);
34.         dest.writeDouble(DivResult);
35.     }
36.
```

```
37.    public static final Parcelable.Creator<AllResult> CREATOR =
38.        new Parcelable.Creator<AllResult>(){
39.        public AllResult createFromParcel(Parcel parcel){
40.            return new AllResult(parcel);
41.        }
42.        public AllResult[] newArray(int size){
43.            return new AllResult[size];
44.        }
45.    };
46. }
```

第 6 行代码说明了 AllResult 类继承于 Parcelable。第 7 行到第 10 行代码用来保存全运算的运算结果。第 12 行是 AllResult 类的基本构造函数。第 19 行是类的构造函数，支持通过 Parcel 对象实例化 AllResult。第 20 行到第 23 行代码是构造函数的读取顺序。第 32 行代码的 writeToParcel() 是"打包"函数，将 AllResult 类内部的数据，按照特定的顺序写入 Parcel 对象，写入的顺序必须与构造函数的读取顺序一致。第 39 行实现了静态公共字段 Creator，用来使用 Parcel 对象构造 AllResult 对象。

在 MathService.java 文件中，增加了用来进行全运算的 ComputAll() 函数，并将运算结果保存在 AllResult 对象中。ComputAll() 函数实现代码如下。

```
1.    @Override
2.    public AllResult ComputeAll(long a, long b) throws RemoteException {
3.        long addRusult = a + b;
4.        long subResult = a - b;
5.        long mulResult = a * b;
6.        double divResult = (double) a / (double)b;
7.        AllResult allResult = new AllResult(addRusult, subResult,
            mulResult, divResult);
8.        return allResult;
9.    }
```

ParcelMathCallerDemo 示例是 ParcelMathServiceDemo 示例中 MathService 服务的调用者。ParcelMathCallerDemo 文件结构如图 7-15 所示。

其中 AllResult.aidl、AllResult.java 和 IMathService.aidl 文件务必与 ParcelMathServiceDemo 示例的三个文件完全一致，否则会出现错误。

第 7 章　Android 后台服务

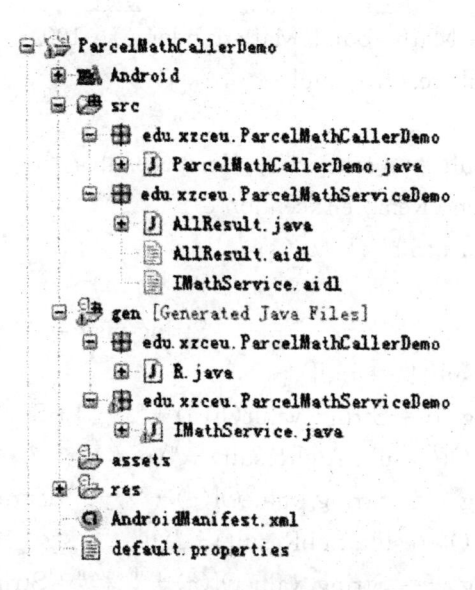

图 7-15　ParcelMathCallerDemo 文件结构

下图 7-16 是 ParcelMathCallerDemo 用户界面。

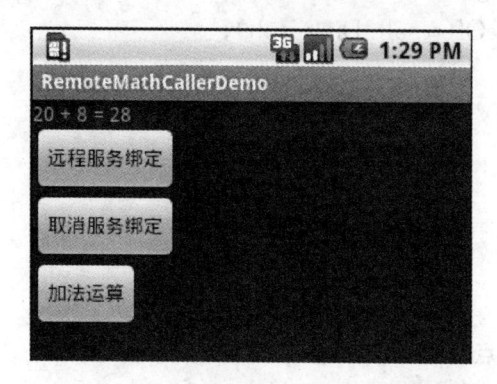

图 7-16　ParcelMathCallerDemo 用户界面

原来的"加法运算"按钮改为了"全运算"按钮，运算结果显示在界面的上方。下面是 ParcelMathCallerDemo.java 文件与 RemoteMathCallerDemo 示例中 RemoteMathCallerDemo.java 文件不同的一段代码，定义了"全运算"按钮的监听函数。随机产生输入值，调用跨进程服务，获取运算结果，并将运算结果显示在用户界面上。

```
1.   computAllButton.setOnClickListener(new View.OnClickListener(){
2.   @Override
3.   public void onClick(View v) {
4.      if (mathService == null){
5.        labelView.setText("未绑定跨进程服务");
6.        return;
7.      }
8.      long a = Math.round(Math.random() * 100);
```

```
9.      long b = Math.round(Math.random()*100);
10.     AllResult result = null;
11.     try {
12.         result = mathService.ComputeAll(a, b);
13.     } catch (RemoteException e) {
14.         e.printStackTrace();
15.     }
16.                     String msg = "";
17.     if (result != null){
18.         msg += String.valueOf(a)+" + "+String.valueOf(b)+" = "+String.valueOf(result.AddResult)+"\n";
19.         msg += String.valueOf(a)+" - "+String.valueOf(b)+" = "+String.valueOf(result.SubResult)+"\n";
20.         msg += String.valueOf(a)+" * "+String.valueOf(b)+" = "+String.valueOf(result.MulResult)+"\n";
21.         msg += String.valueOf(a)+" / "+String.valueOf(b)+" = "+String.valueOf(result.DivResult);
22.     }
23.     labelView.setText(msg);
24.     }
25. });
```

习题与思考题

1. 简述 Service 的特点。
2. 简述本地服务的实现。
3. 简述远程服务的实现。

第 8 章 对话框与提示信息

☆8.1 对话框
☆8.2 提示信息 Toast
☆8.3 温馨信息 Notification

本章学习目标：掌握对话框设计方法；了解提示信息的设计与使用方法。

8.1 对话框

对话框是一种显示于 Activity 之上的界面元素，是作为 Activity 的一部分被创建和显示的。常用的对话框种类有：提示对话框 AlertDialog，进度对话框 ProgressDialog，日期选择对话框 DatePickerDialog，时间选择对话框 TimePickerDialog。如图 8-1 所示。

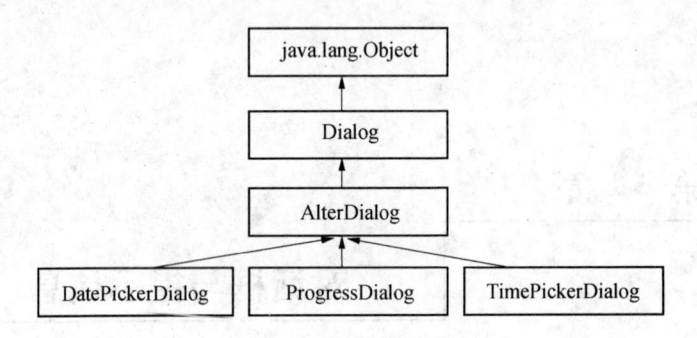

图 8-1 常用对话框类型

创建 AlertDialog 对话框的主要步骤：获得 AlertDialog 静态内部类 Builder 对象并由该类创建对话框，通过 Buidler 对象设置对话框标题、按钮及其将要响应事件，调用 Builder 的 create()方法创建对话框，调用 AlertDialog 的 show()方法显示对话框。

它提供的方法主要有以下种类。setTitle()：设置对话框标题。setIcon()：设置对话框图标。setMessage()：设置对话框提示信息。setItems()：设置对话框要显示的一个列表。setSingleChoiceItems()：设置对话框显示一个单选的 List。setMultiChoiceItems()：设置对话框显示一系列的复选框。setPositiveButton()：给对话框添加 Yes 按钮。setNegativeButton()：给对话框添加 No 按钮。setView()：给对话框设置自定义样式。create()：创建对话框。show() 和 showDialog()：显示对话框。onCreateDialog()：创建对话框的实现。onPrepareDialog()：更改已有对话框时调用。

8.1.1 创建简单的提示对话框

当创建对话框时，首先需要重写 onCreate()方法，可通过 setTitle 设置标题、setContentView()设置内容等。

> 1. finalViewmyviewondialog=usingdialoglayoutxml.inflate(R.layout.dialogshow, null); //设定的布局
> 2. AlertDialogmydialoginstance=newAlertDialog.Builder(ShowDialog_Activity.this)

设置属性，包括标题、按钮和图标等。

> 1. setIcon(R.drawable.icon)//图标，显示在对话框标题左侧
> 2. setTitle("用户登录界面")//对话框标题

> 3. setView(myviewondialog) // 参数为上面定义的 View 实例名,显示 R. layout. dialogshow. xml 布局文件

简单提示对话框案例如下。

```
1. package com. AlterDlgDemo;
2. import android. app. Activity;
3. import android. app. AlertDialog;
4. import android. os. Bundle;
5. public class AlterDlgDemo_MainActivity extends Activity {
6.     @Override
7.     public void onCreate(Bundle savedInstanceState) {
8.         super. onCreate(savedInstanceState);
9.         setContentView(R. layout. main );
10.        AlertDialog. Builder my_ADialog = new AlertDialog. Builder( this );// 新建 AlertDialog. Builder 对象
11.        my_ADialog. setTitle( "Android 提示 " );// 设置标题
12.        my_ADialog. setMessage( "这个是 AlertDialog 提示对话框!!" );// 设置显示消息
13.        my_ADialog. show();// 显示
14.    }
15. }
```

运行效果如图 8-2 所示。

图 8-2 简单提示对话框

8.1.2 创建具有简单界面的提示对话框

创建提示对话框包括：在相应的工程中修改 res\layout\main.xml 文件，添加 Button 按钮，并指定显示模式（如按钮是否为包裹住文字内容）、id、文字等；为即将添加的对话框设计布局；在 Java 代码中随着触发事件而弹出相应的 AlterDialog 对话框，其具体方法如下。

通过 setContentView(R.layout.main) 语句设定使用默认的布局文件。定义一个 Button，这个 Button 通过 findViewById() 和在 main.xml 中已经设定好的按钮建立了联系，因此显示在默认布局中的按钮就是在 main.xml 中定义好的按钮。通过 OnClickListener，侦听此按钮的被单击事件。如果按钮被单击，则定义一个 LayoutInflater 类的实例。LayoutInflater 类的作用类似于 findViewById()，不同点是 LayoutInflater 是用来找 layout 下 xml 布局文件并且实例化，而 findViewById() 是找具体 xml 下具体 widget 控件（如 Button、TextView 等）。设定这个对话框的标题并设定对话框中的按钮以及相应按钮对应的单击事件。通过 setView(已经定义的 View 实例名)，显示 R.layout.dialogshow.xml 这个布局文件。通过 onCreate() 创建 Activity。通过调用对话框实例的 show() 方法显示这个对话框。

运行效果如图 8-3(a)、(b)、(c)所示。

(a)　　　　　　　　　　　(b)

(c)

图 8-3　具有简单界面的提示对话框

8.1.3 创建多种不同类型的提示对话框

在 string.xml 中设置相关的字符变量,在 array.xml 中设置选项。在布局文件 main.xml 中定义一个 EditText 及三个 Button,分别用于显示相应的文本信息,以及用于打开相应的对话框。在 Activity 代码中通过定义 Button 实例来引用在布局文件中定义的按钮。通过侦听按钮被单击的事件来调用 showDialog(对话框 ID)来打开相应对话框。重写 onCreateDialog 方法,在其中根据不同的形参 ID 调用不同的处理逻辑,如设置其图标、标题、添加不同的按钮并设定按钮对应的处理逻辑等。最后,通过 create()方法的调用生成该对话框。运行效果如图 8-4 所示。

图 8-4 多种不同类型的提示对话框

8.2 提示信息 Toast

Toast 是 Android 中用来显示提示信息的一种机制。和对话框 Dialog 不一样的是,Toast 没有焦点且显示的时间有限,不会打断用户当前的操作,信息在 floating view 呈现,然后会自动消失。创建 Toast 的一般步骤是:首先调用 Toast 的静态方法 makeText()或 make()添加显示文本和时长,即 Toast.makeText(getApplicationContext(),"显示文本",显示时长);其次调用 Toast 的 show()显示提示信息。

Toast 案例的代码如下。

```
1.  public class ToastDemo_MainActivity extends Activity {
2.    private Button mButton,mButton2 ;//实例化两个按钮
3.    private EditText mEditText ;//实例化一个 EditText
4.      @Override
```

```
5.    public void onCreate(Bundle savedInstanceState) {
6.        super.onCreate(savedInstanceState);
7.        setContentView(R.layout.main);//采用main布局
8.        mButton=(Button)findViewById(R.id.mybtn);//找到id号为指定内容按钮并和自定义的实例关联
9.        mButton2=(Button)findViewById(R.id.mybtn2);//找id号为指定内容按钮并和自定义的实例关联
10.       mEditText=(EditText)findViewById(R.id.myet);//找id号为指定内容控件并和自定义的实例关联
11.       mButton.setOnClickListener( new Button.OnClickListener(){//侦听按钮1被按下的动作
12.           public void onClick(View v) {
13.               Toast.makeText(getApplicationContext(),"您的愿望",Toast.LENGTH_SHORT).show();//短时
14.               Editable Str;
15.               Str=mEditText.getText();//得到用户输入的内容
16.               CharSequence string2 = getString(R.string.yourwish );//用CharSequence类getString()方法从XML中获取String
17.               CharSequence string3=getString(R.string.send );//用CharSequence类getString()方法从XML中获取String
18.               Toast.makeText(ToastDemo_MainActivity.this, string2+Str.toString()+string3,Toast.LENGTH_LONG).show();//用makeText()方式产生Toast信息,时长为较长
19.               mEditText.setText("");//清空EditText
20.           }
21.       });
22.       mButton2.setOnClickListener(new Button.OnClickListener(){//侦听按钮2被按下的动作
23.           public void onClick(View v) {
               Toast.makeText(getApplicationContext(),"我的表情",Toast.LENGTH_SHORT).show();//短时
24.               Toast toast = new Toast(getApplicationContext());//toast实例化
25.               ImageView myview = new ImageView(getApplicationContext());//实例化ImageView
26.               myview.setImageResource(R.drawable.image4);//和指定的图片关联
27.               toast.setView(myview);//将toast实例和图片实例关联
28.               toast.show();//显示toast
29.           }
```

```
30.     });
31.   }
32. }
```

运行效果如图 8-5 所示。

图 8-5　Toast 示例

8.3　温馨信息 Notification

和 Toast 不同,温馨提示 Notification 是 Android 提供的在状态栏的提醒机制。同样,它也不会打断用户当前的操作,而且支持更复杂的点击事件响应。Notification 使用 NotificationManager 来管理创建 Notification 的步骤。首先,得到 NotificationManager 的引用;第二,初始化一个 Notification;第三,设置 Notification 的参数;第四,显示 Notification。

Notification 案例的代码如下。

```
1. public class NofiticationDemo_MainActivity extends Activity {
2.   private NotificationManager mNotificationManager;
3.   private int SIMPLE_NOTFICATION_ID;
4.   @Override
5.   public void onCreate(Bundle savedInstanceState) {
6.     super.onCreate(savedInstanceState);
7.     setContentView(R.layout.main);//采用设定的布局,在其中定义了两
个按钮
```

8.   mNotificationManager = (NotificationManager)getSystemService(NOTIFICATION_SERVICE); //创建 NotificationManager 对象,负责"发出"与"取消" Notification。
　　　　　　//下面创建 Notification,参数依次为 icon 的资源 id、在状态栏上展示的滚动信息、时间。
9.   final Notification notifyDetails = new Notification(R.drawable.image5,"滚动提示信息",System.currentTimeMillis());
10.   Button start = (Button)findViewById(R.id.notifyButton); //start 对应显示提示的按钮
11.   Button cancel = (Button)findViewById(R.id.cancelButton); //cancel 对应关闭程序的按钮
12.   start.setOnClickListener(new OnClickListener() {　　//"显示提示"按钮对应的点击事件
13.   public void onClick(View v) {
14.   Context context = getApplicationContext();
15.   CharSequence contentTitle = "您好";
16.   CharSequence contentText = "您选中了显示提示";
17.   setTitle("您选中了显示提示"); //设置 Activity 显示的标题
18.   Intent notifyIntent = new Intent(android.content.Intent.ACTION_VIEW);
　　　　　　//PendingIntent 为 Intent 的包装,这里是启动 Intent 的描述
　　　　　　// PendingIntent.getActivity 返回的 PendingIntent 表示此 PendingIntent 实例中的 Intent 是用于启动 Activity 的 Intent。PendingIntent.getActivity 的参数依次为:Context,发送者的请求码(可以填 0),用于系统发送的 Intent,标志位。
19.   PendingIntent intent = PendingIntent.getActivity(NofiticationDemo_MainActivity.this, 0, notifyIntent, android.content.Intent.FLAG_ACTIVITY_NEW_TASK);
20.   notifyDetails.setLatestEventInfo(context, contentTitle, contentText, intent);
21.   mNotificationManager.notify(SIMPLE_NOTFICATION_ID, notifyDetails);
22.   }
23.   });
24.   cancel.setOnClickListener(new OnClickListener() {　　//"关闭程序"按钮对应的点击事件
25.   public void onClick(View v) {
26.   mNotificationManager.cancel(SIMPLE_NOTFICATION_ID);
27.   finish(); //退出应用程序

28.   }
29.   });
30. }

运行效果如图 8-6 所示。

图 8-6 Nofitication 示例

习题与思考题

1. 简述如何创建提示信息 Toast 的方法。
2. 简述如何创建温馨信息 Notification 的方法。

第 9 章

Android 桌面组件

☆9.1　AppWidget 框架类
☆9.2　如何使用 Widget
☆9.3　Demo 讲解

本章学习目标：了解 App-Widget 框架类；了解在 Android 如何使用 Widget；通过一个 Demo 讲解了解 Widget 的实现。

9.1 AppWidget 框架类

AppWidgetProvider：继承自 BroadcastRecevier，在 AppWidget 应用 update、enable、disable 和 delete 时接收通知。其中，onUpdate、onReceive 是最常用到的方法，它们接收更新通知。

AppWidgetProvderInfo：描述 AppWidget 的大小、更新频率和初始界面等信息，以 XML 文件形式存在于应用的 res/xml/目录下。

AppWidgetManger：负责管理 AppWidget，向 AppwidgetProvider 发送通知。

RemoteViews：一个可以在其他应用进程中运行的类，向 AppWidgetProvider 发送通知。

9.2 如何使用 Widget

长按主界面之后弹出一个对话框，里面就有 android 内置的一些桌面组件。

AppWidget 框架的主要类介绍如下。

（1）AppWidgetManger 类

bindAppWidgetId(int appWidgetId，ComponentName provider)通过给定的 ComponentName 绑定 appWidgetId。

getAppWidgetIds(ComponentName provider)通过给定的 ComponentName 获取 AppWidgetId。

getAppWidgetInfo(int appWidgetId)通过 AppWidgetId 获取 AppWidget 信息。

getInstalledProviders()返回一个 List<AppWidgetProviderInfo>的信息。

getInstance(Context context)获取 AppWidgetManger 实例使用的上下文对象。

updateAppWidget(int[] appWidgetIds，RemoteViews views)通过 appWidgetId 对传进来的 RemoteView 进行修改，并重新刷新 AppWidget 组件。

updateAppWidget(ComponentName provider，RemoteViews views)通过 ComponentName 对传进来的 RemoeteView 进行修改，并重新刷新 AppWidget 组件。

updateAppWidget(int appWidgetId，RemoteViews views)通过 appWidgetId 对传进来的 RemoteView 进行修改，并重新刷新 AppWidget 组件。

（2）继承自 AppWidgetProvider 可实现的方法

继承自 AppWidgetProvider 可实现的方法为如下：

① onDeleted(Context context，int[] appWidgetIds)

② onDisabled(Context context)

③ onEnabled(Context context)

④ onReceive(Context context，Intent intent)

注意：因为 AppWidgetProvider 是继承自 BroadcastReceiver 所以可以重写 onRecevie 方

法,当然必须在后台注册 Receiver。

⑤ onUpdate(Context context, AppWidgetManager appWidgetManager, int[] appWidgetIds)。

9.3 Demo 讲解

下面的实例供大家练习时做参考,要求效果如下:在布局中放一个 TextView 做桌面组件,然后设置 TextView 的 Clickable="true" 使其有点击的功能,然后点击它时改变它的字体,再点击时变回来,详细操作流程如下。

① 新建 AppWidgetProvderInfo
② 写一个类继承自 AppWidgetProvider
③ 后台注册 Receiver
④ 使 AppWidget 组件支持点击事件
⑤ 使 TextView 在两种文本间来回跳转

(1) 新建 AppWidgetProvderInfo

代码如下:

```
<? xml version="1.0" encoding="UTF-8"? >
<appwidget-provider
xmlns:android="http://schemas.android.com/apk/res/android"
    android:minWidth="60dp"
    android:minHeight="30dp"
    android:updatePeriodMillis="86400000"

    android:initialLayout="@layout/main">
</appwidget-provider>
```

注意:上文说过 AppWidgetProvderInfo 是在 res/xml 的文件形式存在的,看参数不难理解,比较重要的是 android:initialLayout="@layout/main",此句为指定桌面组件的布局文件。

(2) 写一个类继承自 AppWidgetProvider

主要代码如下:

```
public class widgetProvider extends AppWidgetProvider
```

并重写两个方法:

```
@Override
    public void onUpdate(Context context, AppWidgetManager appWidgetManager,
            int[] appWidgetIds){}


@Override
    public void onReceive(Context context, Intent intent){}
```

注意:onUpdate 为组件在桌面上生成时调用,并更新组件 UI,onReceiver 为接收广播时调用更新 UI,一般这两个方法是比较常用的。

(3) 后台注册 Receiver

后台配置文件代码如下:

```
<receiver android:name=".widgetProvider">
    <meta-data android:name="android.appwidget.provider"
        android:resource="@xml/appwidget_provider"></meta-data>
    <intent-filter>
        <action android:name="com.terry.action.widget.click"></action>
        <action android:name="android.appwidget.action.APPWIDGET_UPDATE" />
    </intent-filter>
</receiver>
```

注意:因为是桌面组件,所以暂时不考虑使用 Activity 界面,当然在实现做项目时,可能会需要点击时跳转到 Activity 应用程序上做操作,典型的案例为 Android 提供的音乐播放器。上面代码中比较重要的是这一句 <meta-data android:name="android.appwidget.provider" android:resource="@xml/appwidget_provider"></meta-data>大意为指定桌面应用程序的 AppWidgetProvderInfo 文件,使其可作其管理文件。

(4) 使 AppWidget 组件支持点击事件

先看代码:

```
public static void updateAppWidget(Context context,
        AppWidgetManager appWidgeManger, int appWidgetId) {
    rv = new RemoteViews(context.getPackageName(), R.layout.main);
    Intent intentClick = new Intent(CLICK_NAME_ACTION);
    PendingIntent pendingIntent = PendingIntent.getBroadcast(context, 0,
        intentClick, 0);
    rv.setOnClickPendingIntent(R.id.TextView01, pendingIntent);
    appWidgeManger.updateAppWidget(appWidgetId, rv);
}
```

此方法为创建组件时 onUpdate 调用的更新 UI 的方法,代码中使用 RemoteView 找到组件的布局文件,同时为其设置广播接收器 CLICK_NAME_ACTION,并且通过 RemoteView 的 setOnClickPendingIntent 方法找到想触发事件的 TextView,为其设置广播。接着:

```
@Override
public void onReceive(Context context, Intent intent) {
    // TODO Auto-generated method stub
    super.onReceive(context, intent);
    if (rv == null) {
        rv = new RemoteViews(context.getPackageName(), R.layout.main);
    }
```

```
            if (intent.getAction().equals(CLICK_NAME_ACTION)) {
                if (uitil.isChange) {
                    rv.setTextViewText(R.id.TextView01, context.getResources()
                        .getString(R.string.load));
                } else {
                    rv.setTextViewText(R.id.TextView01, context.getResources()
                        .getString(R.string.change));
                }
                Toast.makeText(context, Boolean.toString(uitil.isChange),
                    Toast.LENGTH_LONG).show();
                uitil.isChange = ! uitil.isChange;


            }
            AppWidgetManager appWidgetManger = AppWidgetManager
                .getInstance(context);
            int[] appIds = appWidgetManger.getAppWidgetIds(new ComponentName(
                context, widgetProvider.class));
            appWidgetManger.updateAppWidget(appIds, rv);
        }
```

在 onReceiver 中通过判断传进来的广播来触发动作。

(5) 使 TextView 在两种文本间来回跳转

使 TextView 能在两种文本间来回跳转是调试过程最久的一个难点，问题出在对 AppWidget 的理解不够深入。AppWidget 的生命周期应该在每接收一次广播执行一次，并作为一个生命周期结束，也就是说在重写的 AppWidgetProvider 类里面声明全局变量做状态判断，每次状态改变 AppWidgetProvider 再接收第二次广播时，即重新初始化也就是说桌件重新实例化了一次 AppWidgetProvider。因为在代码中加入一个 boolean 值初始化为 true，观察调试看到每次进入都为 TRUE，故在设置桌面组件时，全局变量把它声明在另外一个实体类用来判断是没问题的，切忌放在本类。代码参考 onReceiver 方法。

代码如下：

```
package com.terry;
import android.app.PendingIntent;
import android.appwidget.AppWidgetManager;
import android.appwidget.AppWidgetProvider;
import android.content.ComponentName;
import android.content.Context;
import android.content.Intent;
import android.widget.RemoteViews;
import android.widget.Toast;

public class widgetProvider extends AppWidgetProvider {
```

```java
        private static final String CLICK_NAME_ACTION = "com.terry.action.widget.click";
        private static RemoteViews rv;
        @Override
        public void onUpdate(Context context, AppWidgetManager appWidgetManager,
                int[] appWidgetIds) {
            // TODO Auto-generated method stub
            final int N = appWidgetIds.length;
            for (int i = 0; i < N; i++) {
                int appWidgetId = appWidgetIds;
                updateAppWidget(context, appWidgetManager, appWidgetId);
            }
        }
        @Override
        public void onReceive(Context context, Intent intent) {
            // TODO Auto-generated method stub
            super.onReceive(context, intent);
            if (rv == null) {
                rv = new RemoteViews(context.getPackageName(), R.layout.main);
            }
            if (intent.getAction().equals(CLICK_NAME_ACTION)) {
                if (uitil.isChange) {
                    rv.setTextViewText(R.id.TextView01, context.getResources()
                        .getString(R.string.load));
                } else {
                    rv.setTextViewText(R.id.TextView01, context.getResources()
                        .getString(R.string.change));
                }
                Toast.makeText(context, Boolean.toString(uitil.isChange),
                    Toast.LENGTH_LONG).show();
                uitil.isChange = ! uitil.isChange;
            }
            AppWidgetManager appWidgetManger = AppWidgetManager
                    .getInstance(context);
            int[] appIds = appWidgetManger.getAppWidgetIds(new ComponentName(
                    context, widgetProvider.class));
            appWidgetManger.updateAppWidget(appIds, rv);
        }

        public static void updateAppWidget(Context context,
```

```
    AppWidgetManager appWidgeManger, int appWidgetId) {
        rv = new RemoteViews(context.getPackageName(), R.layout.main);
        Intent intentClick = new Intent(CLICK_NAME_ACTION);
        PendingIntent pendingIntent = PendingIntent.getBroadcast(context, 0,
                intentClick, 0);
        rv.setOnClickPendingIntent(R.id.TextView01, pendingIntent);
        appWidgeManger.updateAppWidget(appWidgetId, rv);
    }
}
```

习题与思考题

1. 简述 AppWidget 框架类。
2. 简述 AppWidgetProvider 可实现的方法。

第 10 章

SQL 基础

☆10.1　SQL 概述
☆10.2　数据定义功能
☆10.3　数据操纵功能
☆10.4　查询功能

本章学习目标：了解 SQL；掌握使用 SQL 创建表、删除表、修改表结构的方法；掌握使用 SQL 查询的方法。

10.1 SQL 概述

SQL 是结构化查询语言 Structured Query Language 的缩写，是关系数据库的标准语言。虽然绝大多数的关系数据库管理系统都支持 SQL，但是它们有各自特有的扩展功能用于自己的系统。VFP 数据库管理系统中除了具有 VFP 命令之外还支持 SQL 命令。由于 SQL 命令利用 Rushmore 技术实现优化处理，一条 SQL 命令可以替代多条 VFP 命令。

SQL 标准于 1986 年 10 月由美国 ANSI 公布，随后 1987 年 6 月国际标准化组织 ISO 将 SQL 定为国际标准，推荐它成为标准关系数据库语言。1990 年，我国颁布了《信息处理系统数据库语言 SQL》，将其定为中国国家标准。

SQL 语言具有如下主要特点：

（1）SQL 是一种一体化语言，包括数据定义、数据查询、数据操纵和数据控制等功能。

（2）SQL 是一种高度非过程化语言，用户只需告诉系统"做什么"，无须描述"怎么做"。

（3）SQL 语言功能极强，但是非常简洁，完成数据定义、数据查询、数据操纵和数据控制等功能只用了为数不多的几个动词，并且 SQL 的语法也非常简单，接近英语自然语言。易学易用是 SQL 最大的特点。

（4）可以直接采用交互方式使用，也可以嵌入程序设计语言在程序方式下使用。

尽管 SQL 功能强大，但是概括起来，可以分成以下几组：

（1）数据操作语言 DML(Data Manipulation Language)：用于检索或者修改数据。

（2）数据定义语言 DDL(Data Definition Language)：用于定义数据的结构，比如创建、修改或者删除数据库对象。

（3）数据控制语言 DCL(Data Control Language)：用于定义数据库用户的权限。

VFP 在 SQL 方面支持数据定义、数据查询和数据操纵功能，但在具体实现方面也存在一些差异，如没有提供数据控制功能，不支持多层嵌套查询等。在 VFP 环境下，SQL 命令既可以在命令窗口直接执行，也可以写入 .PRG 文件中执行。

10.2 数据定义功能

SQL 的数据定义命令主要包括 CREATE（定义表结构）、DROP（删除表）和 ALTER（修改表结构）等命令。

10.2.1 定义表结构

【语法格式】
CREATE TABLE|DBF <表名1> [NAME <长表名>] [FREE]
(<字段名1> <类型>(<宽度>[,<小数位数>])[NULL|NOT NULL]

[CHECK <约束条件1>[ERROR <出错提示信息1>]]
[DEFAULT <默认值1>]
[PRIMARY KEY|UNIQUE]
[REFERENCES <表名2> [TAG <索引标识1>]]
[,<字段名2> <类型>(<宽度>[,<小数位数>]) [NULL|NOT NULL]
[CHECK <约束条件2>[ERROR <出错提示信息2>]]
[DEFAULT <默认值2>]
[,PRIMARY KEY <索引表达式2> TAG <索引标识2>
[,UNIQUE <索引表达式3> TAG <索引标识3>]
[,FOREIGN KEY <索引表达式4> TAG <索引标识4>]REFERENCES <表名3>
[TAG <索引标识5>]] …)

【功能】

创建自由表或数据库表、建立索引、定义域完整性和表间的联系。CREATE TABLE 命令可以实现表设计器的全部功能；关键字 TABLE 和 DBF 的作用相同，TABLE 是标准 SQL 的关键字，DBF 是 VFP 的关键字；SQL CREATE 命令建立的表自动在最低可用工作区打开，新表的打开方式为独占方式，不受 SET EXCLUSIVE 设置影响。

【说明】

FREE：选择该选项，则新建表不添加到当前数据库，成为一个自由表，那么命令后面的 NAME、CHECK、DEFAULT、FOREIGN KEY、PRIMARY KEY 和 REFERENCE 等选项均不能选用。该项缺省，说明建立一个数据库表。

(<字段名1> <类型>(<宽度>[,<小数位数>])[NULL|NOT NULL]：定义表中的字段，选项 NULL|NOT NULL 用于声明是否允许该字段值为空值。字段类型需要用一个特定的字母表示。

CHECK <约束条件1>[ERROR <出错提示信息1>][DEFAULT <默认值1>]：定义字段有效性规则。

PRIMARY KEY：将字段定义为主索引。

UNIQUE：将字段定义为候选索引，注意这里 UNIQUE 不是唯一索引。

PRIMARY KEY <索引表达式2> TAG <索引标识2>|UNIQUE <索引表达式3> TAG <索引标识3>：按指定的索引表达式建立主索引或候选索引。

FOREIGN KEY <索引表达式4> TAG <索引标识4>：按索引表达式4建立普通索引。

REFERENCES <表名3>[TAG <索引标识5>]：与表3建立永久关联。

例1. 利用 SQL 命令建立学生成绩管理数据库，该数据库包含学生信息表、课程信息表和成信息表。

步骤1：建立学生成绩管理数据库。CREATE DATABASE 学生成绩管理

步骤2：建立学生信息表。CREATE TABLE 学生信息；

(学号 C(8) PRIMARY KEY；

,姓名 C(8)；

,性别 C(2) CHECK 性别="男" OR 性别="女"；

ERROR [只能输入"男"或"女"] DEFAULT "男"；

,出生日期 D;
,入学成绩 N(5,1);
,四级通过否 L;
,计算机等级考试 C(4) NULL;
,备注 M)
步骤3:建立课程信息表。CREATE TABLE 课程信息(课程号 C(2) PRIMARY KEY,课程名 C(10))
步骤4:建立成绩信息表。CREATE TABLE 成绩信息;
(学号 C(8),课程号 C(2),成绩 N(5,1),;
PRIMARY KEY 学号+课程号 TAG 学号课程,;
FOREIGN KEY 学号 TAG 学号 REFERENCES 学生信息)

10.2.2 修改表结构

修改表结构命令 ALTER TABLE 有三种不同的语法格式。
【语法格式1】
ALTER TABLE <表名1>
ADD|ALTER [COLUMN] <字段名> <字段类型>(<宽度>[,<小数位数>])
[NULL|NOT NULL]
[CHECK <约束条件1>[ERROR <出错提示信息1>]]
[DEFAULT <默认值>]
[PRIMARY KEY|UNIQUE]
[REFERENCES <表名2> [TAG <索引标识1>]]
【功能】对指定表添加(ADD)新字段或修改(ALTER)已有的字段、建立主索引、候选索引和普通索引,建立表间的永久关联。
【说明】命令中各子句的含义与 CREATE TABLE 命令类似。
例2. 为课程信息表添加两个字段:学分,N,2;学时,N,2,并要求它们的值不得小于或等于0。为学生信息表中的四级通过否字段设置默认值.F.。
ALTER TABLE 课程信息 ADD 学分 N(2) CHECK 学分>0
ALTER TABLE 课程信息 ADD 学时 N(2) CHECK 学时>0
ALTER TABLE 学生信息 ALTER 四级通过否 L DEFAULT.
例3. 在成绩信息表中将课程号字段定义为索引标识为"课程号的普通索引,并引用课程信息表中的主索引"课程号"与课程信息表建立联系。
ALTER TABLE 成绩信息;
ADD FOREIGN KEY 课程号 TAG 课程号 REFERENCES 课程信息
注意:REFERENCES 子句由子表向父表建立永久关联。
【语法格式2】
ALTER TABLE <表名> ALTER [COLUMN] <字段名> [NULL|NOT NULL]
[SET DEFAULT <默认值>][SET CHECK <约束条件>[ERROR <出错提示信息>]]
[DROP DEFAULT][DROP CHECK]

【说明】SET DEFAULT <默认值>：设置指定字段的默认值；SET CHECK <约束条件>[ERROR <出错提示信息>]：设置指定字段的有效性规则和出错提示信息；DROP DEFAULT 或 DROP CHECK：删除指定字段的默认值或有效性规则。

【功能】对指定表中指定字段的有效性规则和默认值进行设置(SET)、修改(SET)和删除(DROP)。

例4. 为学生信息表中的入学成绩字段设置有效性规则。

ALTER TABLE 学生信息

ALTER 入学成绩 SET CHECK

入学成绩>0

例5. 删除学生信息表中的入学成绩字段的有效性规则。

ALTER TABLE 学生信息

ALTER

入学成绩　DROP CHECK

【语法格式3】

ALTER TABLE <表名1>

[DROP [COLUMN] <字段名>]

[SET CHECK <约束条件> [ERROR <默认值>]]

[DROP CHECK]

[ADD PRIMARY KEY <索引表达式1> TAG <索引标识1>]

[DROP PRIMARY KEY]

[ADD UNIQUE <索引表达式2> [TAG <索引标识2>]]

[DROP UNIQUE TAG <索引标识3>]

[ADD FOREIGN KEY [<索引表达式3>] TAG <索引标识4>

REFERENCES <表名2> [TAG <索引标识>]]

[DROP FOREIGN KEY TAG <索引标识> [SAVE]]

[RENAME COLUMN <源字段名> TO <目标字段名>]

【功能】对指定的表进行删除字段（DROP COLUMN）、修改字段名（RENAME COLUMN），还可以修改指定表的完整性规则。

【说明】DROP [COLUMN] <字段名>：删除指定字段；

DROP CHECK：删除指定字段的有效性规则；

DROP PRIMARY KEY 或 DROP UNIQUE TAG <索引标识3>或 DROP FOREIGN KEY TAG <索引标识>：删除指定表的主索引或候选索引或普通索引；

SET CHECK <约束条件> [ERROR <默认值>]：设置指定字段的有效性规则和默认值；

RENAME COLUMN <源字段名> TO <目标字段名>：修改指定的字段名。

例6. 将学生信息表中的"出生日期"字段名改为"出生年月"。

ALTER TABLE 学生信息 RENAME COLUMN 出生日期 TO 出生年月。

例7. 删除学生信息表中的"备注"字段。

ALTER TABLE 学生信息 DROP COLUMN 备注。

例 8. 将课程信息表中的课程名定义为候选索引,索引表达式是课程名。
ALTER TABLE 课程信息 ADD UNIQUE 课程名。
例 9. 删除课程信息表中的候选索引课程名。
ALTER TABLE 课程信息 DROP UNIQUE TAG 课程名。

10.2.3 删除表

【语法格式】DROP TABLE <表名>
【功能】删除表文件。
【说明】如果指定表文件是当前数据库中的表,则本命令不仅从磁盘上删除指定的.dbf 文件,同时删除该表在数据库文件(.dbc)中有关的信息,否则仅删除相应的磁盘文件。因此,为了避免出现错误提示,建议最好对当前数据库中的表执行本命令,或在项目管理器中选择数据库执行移去操作。
例 10. 删除教师信息表。
DROP TABLE 教师信息。

10.3 数据操纵功能

SQL 的数据操纵命令主要包括 INSRET(插入记录)、UPDATE(修改记录)、DELETE(删除记录)和 SELECT(查询)命令。

10.3.1 插入记录

【语法格式】
INSERT INTO <表名>
[(<字段名 1>[,<字段名 2>,…])]
VALUES(<表达式 1>[,<表达式 2>,…])
【功能】向 INTO 短语指定的表尾插入一条新记录,其值为 VALUES 后面表达式的值。
例 11. 向学生信息表中插入记录。
INSERT INTO 学生信息;
VALUES("00000001","张三","男",{^1982-08-09},580,.T.,"二级")
INSERT INTO 学生信息(学号,姓名,入学成绩) VALUES("00000002","李四",560)

第一条命令插入完整记录,这时字段名可以缺省;第二条命令中只给出插入记录的部分数据,需要给出相应的字段名加以说明。

10.3.2 更新记录

【语法格式】
UPDATE <表名>
SET <字段名1>=<表达式1> [,<字段名2>=<表达式2>…]
[WHERE <条件>]
【功能】修改指定表中指定字段的记录数据。
【说明】UPDATE 命令中,如果 WHERE 子句缺省,更新所有记录指定字段的值;否则只更新满足条件的记录的值。

例12. 将所有"英语"课的成绩置为 0。
UPDATE 成绩 SET 成绩=0 WHERE;
课程号 IN(SELECT 课程号 FROM 课程 WHERE 课程名="英语")

10.3.3 删除记录

【语法格式】
DELETE FROM <表名> [WHERE <条件>]
【功能】
删除指定表中的记录。
【说明】
(1) WHERE 子句缺省,删除所有记录;否则只删除满足条件的记录。
(2) 本命令完成逻辑删除,物理删除需要使用 PACK 命令;恢复需要使用 RECALL 命令。

例13. 逻辑删除学生信息表中的所有记录。
DELETE FROM 学生信息。

10.4 查询功能

SQL 的核心是查询,查询命令称为 SELECT 命令,它的语法格式如下。
【语法格式】
SELECT [ALL|DISTINCT]
[<别名>.]<字段名1> [AS <显示列名>] [,<别名>.]<字段名2> [AS <显示列名>…]
FROM [FORCE][<数据库名>!]<表名>[[AS] <本地别名>]
[[INNER|LEFT[OUTER]|RIGHT [OUTER]|FULL[OUTER]
JOIN <数据库名>!<表名>[[AS] <本地别名>]
[ON <连接条件>…]
[[INTO <目标>]

|TO FILE ＜文件名＞[ADDTIVE]
|TO PRINTER [PROMPT]
【功能】查询。
【说明】SELECT 命令具有多个选项,但是它的基本结构是:
SELECT ＜字段名表＞
FROM ＜表名＞
[WHERE ＜筛选条件＞]
其中:SELECT 指定输出的字段,FROM 指定数据来源,WHERE 指定查询条件。

10.4.1　简单查询

所谓简单查询就是基于上述基本结构,至多具有简单查询条件的查询。
【语法格式】
SELECT [ALL|DISTINCT]
＜字段名 1＞ [AS ＜显示列名＞] [,＜字段名 2＞ [AS ＜显示列名＞…]]
FROM ＜表名＞
[WHERE ＜筛选条件＞]
【说明】
SELECT ＜字段名表＞:指定查询结果输出的字段;
ALL|DISTINCT:ALL 表示输出查询结果中所有记录,DISTINCT 则保证在查询结果中相对于某个字段值的记录是唯一的;
AS ＜显示列名＞:在输出时不希望使用原来的字段名,可以用＜显示列名＞重新设置;
FROM ＜表名＞:FROM 子句后面可以是单个或多个表名,或者视图名,说明要查询的数据来源。选择工作区与打开指定的表均由 VFP 系统安排;
WHERE ＜筛选条件＞:说明查询条件,即选择元组的条件。
例 14. 检索学生关系中所有元组。
SELECT ＊ FROM 学生信息
例 15. 显示所有学生的学号和姓名。
SELECT 学号,姓名 FROM 学生信息,如表 10-1 所示。

表 10-1　学生信息列表

学号	姓名	性别	出生年月	入学成绩	四级通过否	计算机等级考试
04001001	尚杰	男	11/20/86	520.5	T	.NULL.
04001002	余芳习	女	12/26/86	513.5	F	二级
04001057	张铁一	男	01/09/86	612.0	T	.NULL.
04002023	陶红莉	女	02/14/85	535.0	F	.NULL.
04002037	皇甫俊	男	10/07/86	480.0	F	一级
04003054	张晓	女	05/13/87	589.0	F	二级
04005001	杨欣	男	01/01/85	540.0	T	.NULL.
04005003	马松	男	04/16/86	560.0	F	.NULL.

例 16. 显示至少选修了一门课程的学生的学号。
SELECT 学号 FROM 成绩信息观察结果不难看出,在查询结果中包含了重复值。根据

题意,本题关注的是在成绩信息表中出现了哪些学生的学号,至于出现几次,并不是我们感兴趣的。加上 DISTINCT 短语去掉重复值:

SELECT 学号 FROM 成绩信息 DISTINCT

例 17. 输出学生关系中所有学生的姓名和截至统计时的年龄。

SELECT 姓名,INT((DATE()-出生年月)/365) AS 年龄 FROM 学生信息

由于表中没有年龄字段,SELECT 子句中第二列的表达式利用出生年月的数据计算出年龄值,AS 短语用来设置显示列名。

例 18. 显示入学成绩在 560 分以上的学生姓名以及入学成绩。

SELECT 姓名,入学成绩 FROM 学生信息 WHERE 入学成绩>=560

【语法小结】

(1) SELECT 查询的结果生成一个只读的临时表,默认输出到屏幕,以浏览方式显示。

(2) <字段名>指定需要输出的列,该项可以是字段名,也可以是函数或表达式。当需要输出所有列时,可以用一个"*"表示,否则用逗号分隔需要输出的字段名。

10.4.2 带特殊运算符的条件查询

(1) BETWEEN…AND 运算符

<字段名>[NOT] BETWEEN <初值> AND <终值>

BETWEEN 运算符用于检测字段的值是否介于指定的范围内。字段名可以是字段名或表达式。BETWEEN 表示的取值范围是连续的。VFP 的函数 BETWEEN(<表达式>,<初值>,<终值>)与之有类似的功能。

(2) IN 运算符

<字段名>[NOT] IN(<表达式 1>[,<表达式 2>…])

IN 运算符用于检测字段的值是否属于表达式集合或子查询。字段名可以是字段名或表达式。IN 表示的取值范围是逗号分隔的若干个值,它表示的取值范围是离散的。VFP 的函数 INLIST(<表达式>,<表达式 1>[,<表达式 2>…])有类似的功能。

(3) LIKE 运算符

<字段名> LIKE <字符表达式>

LIKE 运算符用于检测字段的值是否与样式字符串匹配。字段名是字符型字段或表达式。字符表达式中可以使用通配符,其中通配符%表示零个或多个字符,通配符_(下划线)表示一个字符。其余字符代表自己。

(4) IS NULL 运算符

<字段名> IS [NOT] NULL

IS NULL 运算符用于检测指定字段的值是否为空(NULL),如果字段值为空,返回真。注意,在检测空值时不能写成字段名=NULL 或字段名!=NULL。

例 19. 检索入学成绩在 510~540 分范围内所有学生的信息。

SELECT * FROM 学生信息 WHERE 入学成绩 BETWEEN 510 AND 540

例 20. 检索入学成绩是 520 和 540 分的所有学生的信息。

SELECT * FROM 学生信息 WHERE 入学成绩 IN(520,540)

例21. 检索所有张姓同学的信息。

SELECT * FROM 学生信息 WHERE 姓名 LIKE "张%"

例22. 检索所有已经通过计算机等级考试的学生的学号、姓名和计算机等级考试信息。

SELECT 学号,姓名,计算机等级考试 FROM 学生信息;

WHERE 计算机等级考试 IS NOT NULL

【语法小结】

(1) BETWEEN <初值> AND <终值>,意即在初值与终值界定的某个区间内。注意这是一个闭区间。

(2) 当 IN 运算中包含的表达式只有一项时,可以用关系运算符"="代替;表达式有多项时,可以用逻辑运算符 OR 连接多个关系运算符"="。

10.4.3 简单的计算查询

SQL SELECT 命令支持各种统计汇总函数。常用的计算函数主要有：COUNT、SUM、AVG、MAX 和 MIN。

例23. 在学生信息表中统计学生人数。

SELECT COUNT(*) FROM 学生信息

例24. 统计已开出的课程门数。

SELECT COUNT(DISTINCT 课程号) FROM 成绩信息

例25. 求入学成绩总和。

SELECT SUM(入学成绩) FROM 学生信息

例26. 求"02"号课程的平均成绩。

SELECT AVG(成绩) FROM 成绩信息 WHERE 课程号="02"

例27. 求"02"号课程的最高分。

SELECT MAX(成绩) FROM 成绩信息 WHERE 课程号="02"

【语法小结】

SELECT 命令使用的常用函数如表 10-2 所示。

表 10-2 常用函数

函 数	功 能
COUNT(<字段名>)	对指定字段的值计算个数
COUNT(*)	计算记录个数
SUM(<字段名>)	计算指定的数值列的和
AVG(<字段名>)	计算指定的数值列的平均值
MAX(<字段名>)	计算指定的字符、日期或数值列中的最大值
MIN(<字段名>)	计算指定的字符、日期或数值列中的最小值

注意:(1) 表中<字段名>可以是字段名,也可以是 SQL 表达式。

(2) 上述函数可以用在 SELECT 短语中对查询结果进行计算,也可以在 HAVING 子句中构造分组筛选条件。

10.4.4 分组与计算查询

计算查询是对整个关系的查询,一次查询只能得出一个计算结果。利用分组计算查询则可以通过一次查询获得多个计算结果。分组查询是通过 GROUP BY 子句实现的。

【语法格式】
GROUP BY <分组关键字 1> [,<分组关键字 2>…]
[HAVING <筛选条件>]

【说明】
<分组关键字>:分组的依据。可以是字段名,也可以是 SQL 函数表达式,还可以是字段序号(从 1 开始);
HAVING:对分组进行筛选的条件。HAVING 总是跟在 GROUPBY 后面,不能单独使用。

例 28. 统计各门课程的平均成绩、最高分、最低分和选课人数。
SELECT 课程号,AVG(成绩) AS 平均成绩,MAX(成绩) AS 最高分,
MIN(成绩) AS 最低分,COUNT(学号) AS 人数 FROM 成绩信息;
GROUP BY 课程号

例 29. 分别统计男、女生入学成绩的最高分。
SELECT 性别,MAX(入学成绩) FROM 学生信息 GROUP BY 性别

例 30. 对选修课程超过三门的学生分别统计选修课程的门数。
SELECT 学号,COUNT(*) AS 选课门数 FROM 成绩信息
GROUP BY 学号 HAVING COUNT(*)>2

【语法小结】
(1) GROUP BY 允许按一个或多个字段分组。
(2) 选择 GROUP BY 选项,则输出的有效数据只含分类关键字和计算函数计算的结果,并且输出行数不一定和表中的记录数一致。一般而言,选用 GROUP BY 子句以后,输出结果中包含的行数少于源表的行数。如果 HAVING 子句缺省,输出结果的行数应该与分组的个数相同。
(3) HAVING 与 WHERE 的区别:WHERE 是对表中所有记录进行筛选,HAVING 是对分组结果进行筛选。
(4) 在分组查询中如果既选用了 WHERE,又选用了 HAVING,执行的顺序是先用 WHERE 限定元组,然后对筛选后的记录按 GROUP BY 指定的分组关键字分组,最后用 HAVING 子句限定分组。

10.4.5 简单的嵌套查询

在有的查询中,需要将另一个 SELECT 查询的结果作为条件。如果在一个 SELECT 命令的 WHERE 子句中出现另一个 SELECT 命令,则这种查询称为嵌套查询。只嵌入一层子查询的 SELECT 命令称为单层嵌套查询,嵌入多于一层子查询的 SELECT 命令称为多层嵌

套查询。SQL 允许多层嵌套，但 VFP 只支持单层嵌套查询。嵌套查询由内向外处理，这样外层查询可以利用内层查询的结果。

例 31. 检索所有选修英语课程的学生的学号。
SELECT 学号 FROM 成绩信息 WHERE；
课程号＝(SELECT 课程号 FROM 课程信息 WHERE 课程＝"英语")

例 32. 检索所有入学成绩高于尚杰的学生的姓名、入学成绩和性别。
SELECT 姓名，入学成绩，性别 FROM 学生信息 WHERE；
入学成绩＞(SELECT 入学成绩 FROM 学生信息 WHERE 姓名＝"尚杰")

10.4.6 使用谓词的嵌套查询

在简单的嵌套查询中，子查询都返回单一的值。如果子查询返回的值有多个，通常在外层查询的 WHERE 子句中使用谓词 ANY(SOME)、ALL、IN 或 EXISTS 加以限制。

【语法格式】
（1）＜字段名＞［NOT］IN(＜子查询＞)
IN 是属于的意思，＜字段名＞指定的字段内容属于子查询中任何一个值，运算结果都为真。＜字段名＞可以是字段名或表达式；
（2）＜字段名＞ ＜比较运算符＞［ANY｜SOME｜ALL］(＜子查询＞)
ANY(SOME)：ANY 和 SOME 是同义词。＜字段名＞指定的字段内容与子查询中任何一个值比较的结果为真时，运算结果为真。＜字段名＞可以是字段名或表达式；
ALL：＜字段名＞指定的字段内容与子查询中所有值比较的结果都为真时，运算结果为真。＜字段名＞可以是字段名或表达式；
（3）［NOT］EXISTS（＜子查询＞）
EXISTS 用来检查在子查询中是否有结果返回，即是否存在相应的元组，有结果返回时，运算结果为真。

例 33. 检索至少选修一门课程的学生的学号和姓名。
SELECT 姓名 FROM 学生信息 WHERE；
学号 IN(SELECT 学号 FROM 成绩信息)

例 34. 检索没有选修任何课程的学生的姓名。
SELECT 姓名 FROM 学生信息 WHERE；
学号 NOT IN(SELECT 学号 FROM 成绩信息)

例 35. 检索选修"02"号课程的学生中成绩最高的学生的学号和成绩。
SELECT 学号，成绩 FROM 成绩信息 WHERE；
课程号＝"02" AND 成绩＞＝ALL(SELE 成绩 FROM 成绩信息 WHERE 课程号＝"02")

例 36. 显示入学成绩高于男生最低入学成绩的女生的学号、姓名和入学成绩。
SELECT 学号，姓名，入学成绩 FROM 学生信息 WHERE；
性别＝"女"AND 入学成绩＞ANY(SELE 入学成绩 FROM 学生信息 WHERE 性别＝"男")

例 37. 检索还没有成绩的课程。
SELECT * FROM 课程信息 WHERE NOT EXISTS;
(SELECT * FROM 成绩信息 WHERE 课程号=课程信息.课程号)

【语法小结】
(1) ANY 和 SOME 是同义词,在进行比较运算时,只要子查询中有一行使结果为真,则结果为真;而 ALL 要求子查询中所有行都使结果为真时,结果才为真。
(2) IN 是属于的意思,等价于 ANY。
(3) EXISTS 本身不具备比较运算功能;IN 本身就具有比较运算的功能。

10.4.7 连接查询

上面的查询大多基于一个表,如果查询涉及两个以上的表,则称之为连接查询。连接查询是关系数据库最主要的查询功能。在实现多表查询时,通常利用公共字段将若干个表两两相连,使它们像一个表一样供查询。SELECT 命令支持多表之间的连接查询,并提供了专门的 JOIN 子句。

【语法格式】
SELECT …
FROM <表名 1> [INNER/LEFT/RIGHT/FULL] JOIN <表名 2>
ON <连接条件>
[WHERE <筛选条件>]

【说明】
JOIN:用来连接左右两个<表名>指定的表;
ON:用来指定连接条件;
INNER/LEFT/RIGHT/FULL:有关概念详见"2. 超链接查询"。

1. 简单的连接查询

例 38. 输出所有学生的成绩单,要求给出学号、课程名和成绩等信息。

【方法一】
SELECT 成绩信息.学号,课程信息.课程名,成绩信息成绩;
FROM 成绩信息 JOIN 课程信息 ON 成绩信息.课程号=课程信息课程号

【方法二】
SELECT 成绩信息.学号,课程信息.课程名,成绩信息成绩;
FROM 成绩信息,课程信息 WHERE (成绩信息.课程号=课程信息.课程号)

例 39. 显示四级已通过并且"01"号课程成绩在 80 分以上的记录。
SELECT 姓名,课程名,成绩 FROM 学生信息 a,成绩信息 b,课程信息 c;
WHERE (a.学号=b.学号 AND b.课程号=c.课程号);
AND 四级通过否 AND 成绩>80 AND b.课程号="01"

2. 超链接查询

VFP 提供的 SELECT 命令中,将连接分为内部连接(Inner Join)和外部连接(Outer Join),外部连接又分为左连接(Left Join)、右连接(Right Join)和全连接(Full Join)。其中 Inner Join 等价于 Join,内部连接就是普通连接,上面的例子都是内部连接。

例 40. 列出专业代码为"002"的同学的成绩,包括姓名、课程号和成绩等信息。

专业代码为"002"的同学有两位,陶红莉选修了两门课程,皇甫俊尚未选修一门课程。所以在成绩表中没有皇甫俊的记录。

内部连接:

SELECT 姓名,课程号,成绩 FROM 学生信息 JOIN 成绩信息;
ON 学生信息.学号=成绩信息.学号;
WHERE SUBS(学生信息.学号,3,3)="002"

左连接:

SELECT 姓名,课程号,成绩 FROM 学生信息 LEFT JOIN 成绩信息;
ON 学生信息.学号=成绩信息.学号;
WHERE SUBS(学生信息.学号,3,3)="002"

【语法小结】

(1) FROM 之后多个关系中包含相同的属性名时,必须用关系前缀指明属性所属的关系,格式:<关系名>.<属性名>或<关系名>-><属性名>。

(2) 如果选择了 JOIN 子句,只能在 ON 子句中指定连接条件,这时 WHERE 子句就只能指定筛选条件,表示在已按连接条件产生的记录中筛选记录。JOIN 子句缺省时,可以在 WHERE 子句中指定连接条件和筛选条件。

(3) 连接查询既可以是两个表的连接,也可以是两个以上表的连接(称多表连接),还可以是一个表自身的连接(称自连接)。

(4) VFP 提供的 SQL SELECT 命令支持超链接查询,主要通过选项 INNER/LEFT/RIGHT/FULL 实现。

10.4.8 排序

SQL SEELECT 允许用户根据需要,将查询结果重新排序后输出。

【语法格式】

ORDER BY <排序关键字 1> [ASC/DESC][,<排序关键字 2>[ASC/DESC]…]
[TOP <数值表达式>[PERCENT]]

【说明】

ASC:表示按升序排序;

DESC:表示按降序排序,ASC|DESC 缺省时默认值是升序;

TOP:必须与 ORDER BY 短语同时使用,它的含义:

① 从第一条记录开始,显示满足条件的前 N 个记录,N 为数值表达式的值;

② 选择 PERCENT 短语时,<数值表达式>表示百分比,它的取值范围 0.01~99.99,指定显示前百分之几的记录;PERCENT 短语缺省时,<数值表达式>表示符合条件的记录数,取值范围是 1~32767 间的整数,指定显示前几个记录。

例 41. 分性别按入学成绩降序输出全体学生记录。

SELECT * FROM 学生信息 ORDER BY 性别,入学成绩 DESC

例 42. 显示入学成绩最高的前三位同学的信息。

SELECT * FROM 学生信息 ORDER BY 入学成绩 DESC TOP 3

例 43. 对所有课程都及格的同学,按平均成绩的降序输出每位同学的学号和平均成绩。
SELECT 学号,AVG(成绩) AS 平均成绩 FROM 成绩信息;
GROUP BY 学号 HAVING MIN(成绩)>=60;
ORDER BY 2 DESC

【语法小结】
(1) ORDER BY 是对最终查询结果排序,不可以在子查询中使用该短语。
(2) 排序关键字可以是字段名,也可以是数字。字段名必须是外层 FROM 子句指定的表中的字段,数字是外层 SELECT 指定的输出列的位置序号。
(3) 在 SQL SELECT 命令中如果不仅选用了 WHERE,还选用了 GROUP BY 和 HAVING,以及 ORDER BY 子句,执行的顺序是:先用 WHERE 指定的条件筛选记录,再对筛选后的记录按 GROUP BY 指定的分组关键字分组,然后用 HAVING 子句指定的条件筛选分组,最后执行 SELECT…ORDER BY 对查询的最终结果进行排序输出。

10.4.9 集合的并运算

将两个 SELECT 命令的查询结果通过集合的并运算合并成一个查询结果。
【语法格式】
[UNION [ALL]<SELECT 命令>]
【说明】
ALL:该项缺省,自动去掉重复记录,否则合并全部结果。
进行并操作的两个 SELECT 命令,必须输出相同的字段个数,并且对应的字段必须具有相同的数据类型和宽度。
例 44. 检索所有通过英语四级考试和计算机等级考试的同学的信息。
SELECT * FROM 学生信息 WHERE 四级通过否;
UNION ALL;
SELECT * FROM 学生信息 WHERE 计算机等级考试 IS NOT NULL
返回

10.4.10 查询结果的重定向输出

SQL SELECT 命令的查询结果默认输出到屏幕,但用户可以根据需要重新指定查询结果的输出去向。
【语法格式】
[INTO <目标>]|[TO FILE <文件名> [ADDITIVE]|TO PRINTER]
【说明】
<目标>有三个选项:
ARRAY <数组名>:将查询结果重定向输出到数组名指定的数组中。
CURSOR <临时表名>:将查询结果重定向输出到一个只读的临时表,查询结束后该表即成为当前表,可以访问,不能修改。关闭文件时系统自动将其删除。
DBF|TABLE <表名>:将查询结果重定向输出到一个永久表中,查询结束后该表即成

为当前表。

TO FILE ＜文件名＞[ADDITIVE]:将查询结果重定向输出到文本文件中,该文件的主名由＜文件名＞指定,扩展名默认为.txt。选择 ADDITIVE 则将结果追加在原文件的尾部,否则覆盖原文件。

TO PRINTER:将查询结果重定向输出到打印机。

例 45. 检索选修了两门课程的学生的信息。

SELECT 学生信息.学号,姓名,成绩 FROM 学生信息 JOIN 成绩信息;
ON 学生信息.学号＝成绩信息.学号 INTO CURSOR TEMP
SELECT * FROM TEMP GROUP BY 学号 HAVING COUNT(*)＝2

例 46. 将查询结果保存到永久表 STUDENTS.DBF 中。

SELECT 学号,姓名 FROM 学生信息 INTO TABLE STUDENTS

例 47. 将查询结果保存到数组 STUDENTS 中。

SELECT 学号,姓名 FROM 学生信息 INTO ARRAY STUDEN

例 48. 将查询结果保存到文本文件 STUDENTS.TXT 中。

SELECT 学号,姓名 FROM 学生信息;
TO FILE STUDENTS ORDER BY 入学成绩 DESC TOP 2

习题与思考题

1. SQL 有哪些特点？使用 SQL 有哪些好处？
2. 自行下载安装 MySQL 等数据库,在命令行界面使用 SQL 语言建立数据库和表,并在表中添加若干学生信息数据,其中包括学生的姓名、性别、年龄、学号、班级、成绩等。
3. 使用 SQL 语句,从习题 2 建立的表中查询全部姓"李"的学生,并将他们的成绩修改成 80。

第11章

数据存储和访问

☆11.1 简单存储
☆11.2 文件存储
☆11.3 数据库存储
☆11.4 数据分享

本章学习目标：掌握 Shared-Preferences 的使用方法；掌握各种文件存储的区别与适用情况；了解 SQLite 数据库的特点和体系结构；掌握 SQLite 数据库的建立和操作方法；理解 ContentPro-vider 的用途和原理；掌握 ContentProvider 的创建与使用方法。

11.1 简单存储

11.1.1 SharedPreferences

SharedPreferences 是一种轻量级的数据保存方式。通过 SharedPreferences 可以将 NVP (Name/Value Pair，名称/值对)保存在 Android 的文件系统中，而且 SharedPreferences 完全屏蔽了对文件系统的操作过程。开发人员仅是通过调用 SharedPreferences 对 NVP 进行保存和读取。

SharedPreferences 不仅能够保存数据，还能够实现不同应用程序间的数据共享。SharedPreferences 支持三种访问模式。

① 私有(MODE_PRIVATE)：仅有创建程序有权限对其进行读取或写入。

② 全局读(MODE_WORLD_READABLE)：不仅创建程序可以对其进行读取或写入，其他应用程序也读取操作的权限，但没有写入操作的权限。

③ 全局写(MODE_WORLD_WRITEABLE)：创建程序和其他程序都可以对其进行写入操作，但没有读取的权限。

在使用 SharedPreferences 前，先定义 SharedPreferences 的访问模式。下面的代码将访问模式定义为私有模式。

```
public static int MODE = MODE_PRIVATE;
```

有的时候需要将 SharedPreferences 的访问模式设定为即可以全局读，也可以全局写，这样就需要将两种模式写成下面的方式。

```
public static int MODE = Context.MODE_WORLD_READABLE + Context.MODE_WORLD_WRITEABLE;
```

定义 SharedPreferences 的名称，这个名称与在 Android 文件系统中保存的文件同名。因此，只要具有相同的 SharedPreferences 名称的 NVP 内容，都会保存在同一个文件中。

```
public static final String PREFERENCE_NAME = "SaveSetting";
```

为了可以使用 SharedPreferences，需要将访问模式和 SharedPreferences 名称作为参数，传递到 getSharedPreferences()函数，并获取到 SharedPreferences 对象。

```
SharedPreferences sharedPreferences = getSharedPreferences(PREFERENCE_NAME, MODE);
```

在获取到 SharedPreferences 对象后,则可以通过 SharedPreferences. Editor 类对 SharedPreferences 进行修改,最后调用 commit()函数保存修改内容。SharedPreferences 广泛支持各种基本数据类型,包括整型、布尔型、浮点型和长型等等。

```
1. SharedPreferences. Editor editor = sharedPreferences. edit();
2. editor. putString("Name", "Tom");
3. editor. putInt("Age", 20);
4. editor. putFloat("Height", );
5. editor. commit();
```

如果需要从已经保存的 SharedPreferences 中读取数据,同样是调用 getSharedPreferences()函数,并在函数的第 1 个参数中指明需要访问的 SharedPreferences 名称,最后通过 get<Type>()函数获取保存在 SharedPreferences 中的 NVP。

get<Type>()函数的第 1 个参数是 NVP 的名称,第 2 个参数是在无法获取到数值的时候使用的缺省值。

```
1. SharedPreferences sharedPreferences = getSharedPreferences(PREFERENCE
   _NAME, MODE);
2. String name = sharedPreferences. getString("Name","Default Name");
3. int age = sharedPreferences. getInt("Age", 20);
4. float height = sharedPreferences. getFloat("Height",);
```

11.1.2 示例

通过 SimplePreferenceDemo 示例介绍具体说明 SharedPreferences 的文件保存位置和保存格式。下图 11-1 是 SimplePreferenceDemo 示例的用户界面。

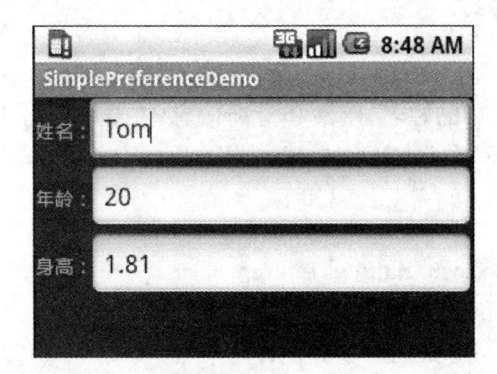

图 11-1 **SimplePreferenceDemo 示例的用户界面**

用户在界面上的输入的信息,将通过 SharedPreferences 在 Activity 关闭时进行保存。当应用程序重新开启时,保存在 SharedPreferences 的信息将被读取出来,并重新呈现在用户界面上。

SimplePreferenceDemo 示例运行后,通过 FileExplorer 查看/data/data 下的数据,Android 为每个应用程序建立了与包同名的目录,用来保存应用程序产生的数据,这些数据包括文件、SharedPreferences 文件和数据库等。SharedPreferences 文件就保存在/data/data/<package name>/shared_prefs 目录下。

在图 11-2 示例中,shared_prefs 目录下生成了一个名为 SaveSetting.xml 的文件。这个文件就是保存 SharedPreferences 的文件,文件大小为 170 字节,在 Linux 下的权限为"-rw-rw-rw"。

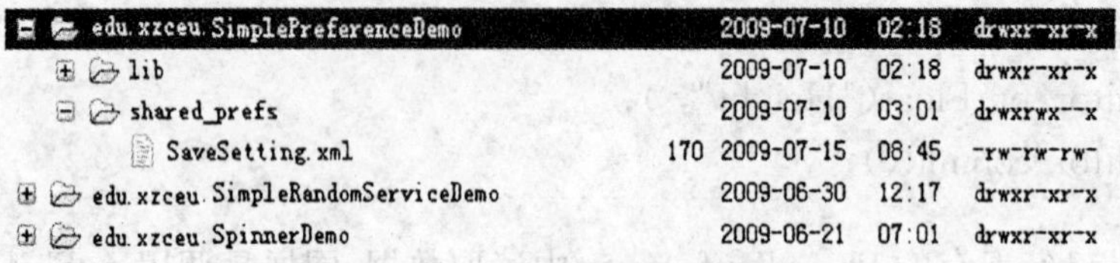

```
edu.xzceu.SimplePreferenceDemo         2009-07-10  02:18   drwxr-xr-x
    lib                                2009-07-10  02:18   drwxr-xr-x
    shared_prefs                       2009-07-10  03:01   drwxrwx--x
        SaveSetting.xml            170 2009-07-15  08:45   -rw-rw-rw-
edu.xzceu.SimpleRandomServiceDemo      2009-06-30  12:17   drwxr-xr-x
edu.xzceu.SpinnerDemo                  2009-06-21  07:01   drwxr-xr-x
```

图 11-2 SaveSetting.xml 的文件

在 Linux 系统中,文件权限分别描述了创建者、同组用户和其他用户对文件的操作限制。x 表示可执行,r 表示可读,w 表示可写,d 表示目录,-表示普通文件。因此,"-rw-rw-rw"表示 SaveSetting.xml 可以被创建者、同组用户和其他用户进行读取和写入操作,但不可执行。产生这样的文件权限与程序人员设定的 SharedPreferences 的访问模式有关,"-rw-rw-rw"的权限是"全局读+全局写"的结果。如果将 SharedPreferences 的访问模式设置为私有,则文件权限将成为"-rw-rw---",表示仅有创建者和同组用户具有读写文件的权限。

SaveSetting.xml 文件是以 XML 格式保存的信息,内容如下。

```
1. <? xml version='1.0' encoding='utf-8' standalone='yes' ? >
2. <map>
3.     <float name="Height" value="1.81" />
4.     <string name="Name">Tom</string>
5.     <int name="Age" value="20" />
6. </map>
```

SimplePreferenceDemo 示例在 onStart()函数中调用 loadSharedPreferences()函数,读取保存在 SharedPreferences 中的姓名、年龄和身高信息,并显示在用户界面上。当 Activity 关闭时,在 onStop()函数调用 saveSharedPreferences(),保存界面上的信息。SimplePreferenceDemo.java 的完整代码如下。

```
1. package edu.xzceu.SimplePreferenceDemo;
2. import android.app.Activity;
3. import android.content.Context;
4. import android.content.SharedPreferences;
5. import android.os.Bundle;
6. import android.widget.EditText;
7. public class SimplePreferenceDemo extends Activity {
8.
```

```
9.     private EditText nameText;
10.    private EditText ageText;
11.    private EditText heightText;
12.    public static final String PREFERENCE_NAME = "SaveSetting";
13.    public static int MODE = Context.MODE_WORLD_READABLE + Context.MODE_WORLD_WRITEABLE;
14.
15.    @Override
16.    public void onCreate(Bundle savedInstanceState) {
17.        super.onCreate(savedInstanceState);
18.        setContentView(R.layout.main);
19.        nameText = (EditText)findViewById(R.id.name);
20.        ageText = (EditText)findViewById(R.id.age);
21.        heightText = (EditText)findViewById(R.id.height);
22.    }
23.
24.    @Override
25.    public void onStart(){
26.        super.onStart();
27.        loadSharedPreferences();
28.    }
29.    @Override
30.    public void onStop(){
31.        super.onStop();
32.        saveSharedPreferences();
33.    }
34.
35.    private void loadSharedPreferences(){
36.        SharedPreferences sharedPreferences = getSharedPreferences(PREFERENCE_NAME, MODE);
37.        String name = sharedPreferences.getString("Name","Tom");
38.        int age = sharedPreferences.getInt("Age", 20);
39.        float height = sharedPreferences.getFloat("Height",);
40.
41.            nameText.setText(name);
42.        ageText.setText(String.valueOf(age));
43.        heightText.setText(String.valueOf(height));
44.    }
45.
```

```
46.    private void saveSharedPreferences(){
47.    SharedPreferences sharedPreferences = getSharedPreferences
    (PREFERENCE_NAME, MODE);
48.    SharedPreferences.Editor editor = sharedPreferences.edit();
49.
50.    editor.putString("Name", nameText.getText().toString());
51.    editor.putInt("Age", Integer.parseInt(ageText.getText().toString()));
52.    editor.putFloat("Height", Float.parseFloat(heightText.getText().
    toString()));
53.    editor.commit();
54.    }
55. }
```

示例 SharePreferenceDemo 将说明如何读取其他应用程序保存的 SharedPreferences 数据。下图 11-3 是 SharePreferenceDemo 示例的用户界面。示例将读取 SimplePreferenceDemo 示例保存的信息，并在程序启动时显示在用户界面上。

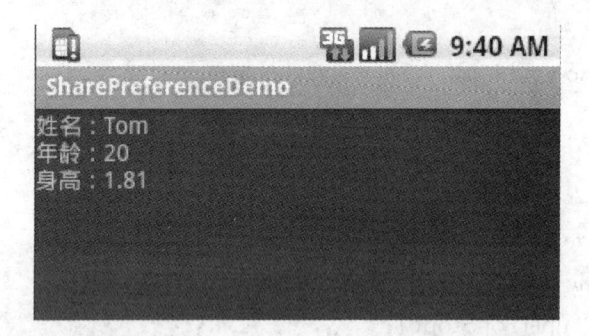

图 11-3　SharePreferenceDemo 示例的用户界面

SharePreferenceDemo 示例的核心代码如下。

```
1. public static final String PREFERENCE_PACKAGE = "edu.xzceu.
    SimplePreferenceDemo";
2. public static final String PREFERENCE_NAME = "SaveSetting";
3. public static int MODE = Context.MODE_WORLD_READABLE +
    Context.MODE_WORLD_WRITEABLE;
4. public void onCreate(Bundle savedInstanceState) {
5.    Context c = null;
6.    try {
7.        c = this.createPackageContext(PREFERENCE_PACKAGE, Context.
    CONTEXT_IGNORE_SECURITY);
8.    } catch (NameNotFoundException e) {
```

```
9.     e.printStackTrace();
10.  }
11.     SharedPreferences sharedPreferences = c.getSharedPreferences
    (PREFERENCE_NAME,MODE);
12.  String name = sharedPreferences.getString("Name","Tom");
13.  int age = sharedPreferences.getInt("Age",20);
14.  float height = sharedPreferences.getFloat("Height",);
15. }
```

第 8 行代码调用了 createPackageContext() 获取到了 SimplePreferenceDemo 示例的 Context。第 8 行代码第 1 个参数是 SimplePreferenceDemo 的包名称,在代码第 1 行进行了定义;第 2 个参数 Context.CONTEXT_IGNORE_SECURIT 表示忽略所有可能产生的安全问题,这段代码可能引发异常,因此必须防止在 try/catch 中。

在代码第 12 行,通过 Context 得到了 SimplePreferenceDemo 示例的 SharedPreferences 对象,同样在 getSharedPreferences() 函数中,需要将正确的 SharedPreferences 名称传递给函数。

访问其他应用程序的 SharedPreferences 必须满足三个条件:

① 共享者需要将 SharedPreferences 的访问模式设置为全局读或全局写。

② 访问者需要知道共享者的包名称和 SharedPreferences 的名称,以通过 Context 获得 SharedPreferences 对象。

③ 访问者需要确切知道每个数据的名称和数据类型,用以正确读取数据。

11.2 文件存储

Android 使用的是基于 Linux 的文件系统,程序开发人员可以建立和访问程序自身的私有文件,也可以访问保存在资源目录中的原始文件和 XML 文件,还可以在 SD 卡等外部存储设备中保存文件。

11.2.1 内部存储

Android 系统允许应用程序创建仅能够自身访问的私有文件,文件保存在设备的内部存储器上,在 Linux 系统下的 /data/data/<package name>/files 目录中。Android 系统不仅支持标准 Java 的 IO 类和方法,还提供了能够简化读写流式文件过程的函数。

下面主要介绍的两个函数:openFileOutput() 和 openFileInput()。

openFileOutput() 函数是为写入数据做准备而打开的应用程序私文件,如果指定的文件不存在,则创建一个新的文件。openFileOutput() 函数的语法格式如下。

```
public FileOutputStream openFileOutput(String name, int mode)
```

第 1 个参数是文件名称,这个参数不可以包含描述路径的斜杠。第 2 个参数是操作模式。

函数的返回值是 FileOutputStream 类型。

Android 系统支持四种文件操作模式,如下表 11-1 所示。

表 11-1　四种文件操作模式

模　式	说　明
MODE_PRIVATE	私有模式,缺陷模式,文件仅能够被文件创建程序访问,或具有相同 UID 的程序访问。
MODE_APPEND	追加模式,如果文件已经存在,则在文件的结尾处添加新数据。
MODE_WORLD_READABLE	全局读模式,允许任何程序读取私有文件。
MODE_WORLD_WRITEABLE	全局写模式,允许任何程序写入私有文件。

使用 openFileOutput()函数建立新文件的示例代码如下。

```
1. String FILE_NAME = "fileDemo.txt";
2. FileOutputStream fos = openFileOutput(FILE_NAME, Context.MODE_PRIVATE)
3. String text = "Some data";
4. fos.write(text.getBytes());
5. fos.flush();
6. fos.close();
```

第 1 行代码定义了建立文件的名称 fileDemo.txt。第 2 行代码使用 openFileOutput()函数以私有模式建立文件。第 4 行代码调用 write()函数将数据写入文件。第 5 行代码调用 flush()函数将所有剩余的数据写入文件。第 6 行代码调用 close()函数关闭 FileOutputStream。

为了提高文件系统的性能,一般调用 write()函数时,如果写入的数据量较小,系统会把数据保存在数据缓冲区中,等数据量累积到一定程度时再一次性的写入文件中。由上可知,在调用 close()函数关闭文件前,务必要调用 flush()函数,将缓冲区内所有的数据写入文件。

openFileInput()函数为读取数据做准备而打开应用程序私文件。openFileInput()函数的语法格式如下。

```
public FileInputStream openFileInput(String name)
```

第 1 个参数也是文件名称,同样不允许包含描述路径的斜杠。

使用 openFileInput()函数打开已有文件的示例代码如下。

```
1. String FILE_NAME = "fileDemo.txt";
2. FileInputStream fis = openFileInput(FILE_NAME);
3.
```

```
4. byte[] readBytes = new byte[fis.available()];
5. while(fis.read(readBytes)!=-1){
6. }
```

上面的两部分代码在实际使用过程中会遇到错误提示,因为文件操作可能会遇到各种问题而最终导致操作失败,因此代码应该使用 try/catch 捕获可能产生的异常。

InternalFileDemo 示例用来演示在内部存储器上进行文件写入和读取。InternalFileDemo 示例用户界面如图 11-4。

图 11-4 InternalFileDemo 示例用户界面

InternalFileDemo 示例的核心代码如下。

```
1. OnClickListener writeButtonListener = new OnClickListener() {
2.     @Override
3.     public void onClick(View v) {
4.         FileOutputStream fos = null;
5.         try {
6.             if (appendBox.isChecked()){
7.                 fos = openFileOutput(FILE_NAME, Context.MODE_APPEND);
8.             }else{
9.                 fos = openFileOutput(FILE_NAME, Context.MODE_PRIVATE);
10.            }
11.            String text = entryText.getText().toString();
12.            fos.write(text.getBytes());
13.            abelView.setText("文件写入成功,写入长度:"+text.length());
14.            entryText.setText("");
```

```
15.              } catch (FileNotFoundException e) {
16. e.printStackTrace();
17.         }
18.         catch (IOException e) {
19.             e.printStackTrace();
20.         }
21.         finally{
22.             if (fos != null){
23.                 try {
24.                     fos.flush();
25.                     fos.close();
26.                 } catch (IOException e) {
27.                     e.printStackTrace();
28.                 }
29.             }
30.         }
31.     }
32. };
33. OnClickListener readButtonListener = new OnClickListener() {
34.     @Override
35.     public void onClick(View v) {
36.         displayView.setText("");
37.         FileInputStream fis = null;
38.         try {
39.             fis = openFileInput(FILE_NAME);
40.             if (fis.available() == 0){
41.                 return;
42.             }
43.             byte[] readBytes = new byte[fis.available()];
44.             while(fis.read(readBytes) != -1){
45.             }
46.             String text = new String(readBytes);
47.             displayView.setText(text);
48.             labelView.setText("文件读取成功,文件长度:"+text.length());
49.         } catch (FileNotFoundException e) {
50. e.printStackTrace();
51.         }
52.         catch (IOException e) {
```

```
53.              e.printStackTrace();
54.         }
55.    }
56. };
```

程序运行后,在/data/data/edu.xzceu.InternalFileDemo/files/目录下,找到了新建立的 fileDemo.txt 文件。fileDemo.txt 文件如图 11-5 所示。

```
edu.hrbeu.InternalFileDemo       2009-07-16  02:28  drwxr-xr-x
    files                        2009-07-16  02:31  drwxrwx--x
        fileDemo.txt         9   2009-07-16  03:24  -rw-rw----
    lib                          2009-07-16  02:28  drwxr-xr-x
```

图 11-5 fileDemo.txt 文件

fileDemo.txt 从文件权限上进行分析,"-rw-rw----"表明文件仅允许文件创建者和同组用户读写,其他用户无权使用。文件的大小为 9 个字节,保存的数据为"Some data"。

11.2.2 外部存储

Android 的外部存储设备指的是 SD 卡(Secure Digital Memory Card),是一种广泛使用于数码设备上的记忆卡。不是所有的 Android 手机都有 SD 卡,但 Android 系统提供了对 SD 卡便捷的访问方法,SD 卡如图 11-6 所示。

SD 卡适用于保存大尺寸的文件或者是一些无须设置访问权限的文件,可以保存录制的大容量的视频文件和音频文件等。SD 卡使用的是 FAT(File Allocation Table)的文件系统,不支持访问模式和权限控制,但可以通过 Linux 文件系统的文件访问权限的控制保证文件的私密性。

Android 模拟器支持 SD 卡,但模拟器中没有缺省的 SD 卡,开发人员须在模拟器中手工添加 SD 卡的映像文件。

图 11-6 SD 卡

使用<Android SDK>/tools 目录下的 mksdcard 工具创建 SD 卡映像文件,命令如下。

```
mksdcard -l SDCARD E:\android\sdcard_file
```

第 1 个参数-l 表示后面的字符串是 SD 卡的标签,这个新建立的 SD 卡的标签是 SDCARD。第 2 个参数表示 SD 卡映像文件的保存位置,上面的命令将映像保存在 E:\android 目录下 sdcard_file 文件中。在 CMD 中执行该命令后,则可在所指定的目录中找到生产的 SD 卡映像文件。

如果希望 Android 模拟器启动时能够自动加载指定的 SD 卡,还需要在模拟器的"运行设置"(Run Configurations)中添加 SD 卡加载命令。SD 卡加载命令中只要指明映像文件位置即可,SD 卡加载命令如图 11-7 所示。

图 11-7 SD 卡命令加载

在模拟器启动后,使用 FileExplorer 向 SD 卡中随意上传一个文件,如果文件上传成功,则表明 SD 卡映像已经成功加载。向 SD 卡中成功上传了一个测试文件 test.txt,文件显示在 /sdcard 目录下,如图 11-8 所示。

图 11-8 上传测试文件 test.txt

编程访问 SD 卡,首先需要检测系统的 /sdcard 目录是否可用。如果不可用,则说明设备中的 SD 卡已经被移除,在 Android 模拟器则表明 SD 卡映像没有被正确加载;如果可用,则直接通过使用标准的 Java.io.File 类进行访问。

将数据保存在 SD 卡,通过"生产随机数列"按钮生产 10 个随机小数通过"写入 SD 卡"按钮将生产的数据保存在 SD 卡的目录下。

SDcardFileDemo 示例说明了如何将数据保存在 SD 卡,下图 11-9 是 SDcardFileDemo 示例的用户界面。

图 11-9 SDcardFileDemo 示例的用户界面

SDcardFileDemo 示例运行后,在每次点击"写入 SD 卡"按钮后,都会在 SD 卡中生产一个新文件,文件名各不相同。SD 卡中生产的文件如图 11-10 所示。

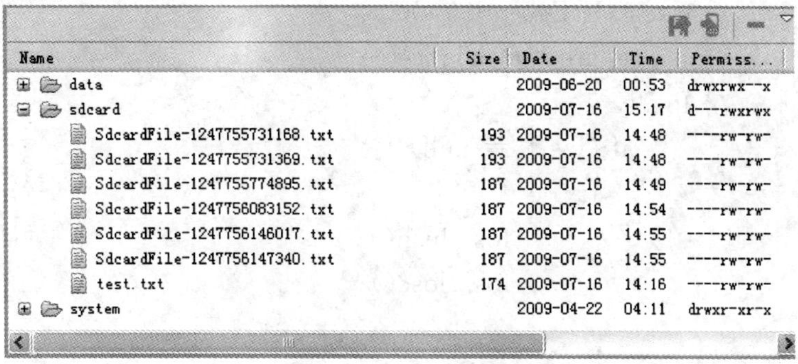

图 11-10 SD 卡中生产的文件

SDcardFileDemo 示例与 InternalFileDemo 示例的核心代码比较相似,不同之处在于:第 7 行代码中添加了/sdcard 目录存在性检查;第 8 行代码使用"绝对目录＋文件名"的形式表示新建立的文件;第 12 行代码写入文件前对文件存在性和可写入性进行检查;第 5 行代码为了保证在 SD 卡中多次写入时文件名不会重复,在文件名中使用了唯一且不重复的标识,这个标识通过调用 System.currentTimeMillis()函数获得,表示从 1970 年 00:00:00 到当前所经过的毫秒数。

下面是 SDcardFileDemo 示例的核心代码。

```
1. private static String randomNumbersString = "";
2. OnClickListener writeButtonListener = new OnClickListener() {
3.     @Override
4.     public void onClick(View v) {
5.         String fileName = "SdcardFile-"+System.currentTimeMillis()+".txt";
6.         File dir = new File("/sdcard/");
7.         if (dir.exists() && dir.canWrite()) {
8.             File newFile = new File(dir.getAbsolutePath() + "/" + fileName);
9.             FileOutputStream fos = null;
10.            try {
11.                newFile.createNewFile();
12.                if (newFile.exists() && newFile.canWrite()) {
13.                    fos = new FileOutputStream(newFile);
14. fos.write(randomNumbersString.getBytes());
15.                    TextView labelView = (TextView)findViewById(R.id.label);
16.                    labelView.setText(fileName + "文件写入 SD 卡");
17.                }
```

```
18.            } catch (IOException e) {
19.                e.printStackTrace();
20.            } finally {
21.                if (fos! = null) {
22.                    try{
23.                        fos.flush();
24.                        fos.close();
25.                    }
26.                    catch (IOException e) { }
27.                }
28.            }
29.        }
30.    }
31. };
```

11.2.3 资源文件

程序开发人员可以将程序开发阶段已经准备好的原始格式文件和 XML 文件分别存放在/res/raw 和/res/xml 目录下,供应用程序在运行时进行访问。原始格式文件可以是任何格式的文件,例如视频格式文件、音频格式文件、图像文件和数据文件等等,在应用程序编译和打包时,/res/raw 目录下的所有文件都会保留原有格式不变。

/res/xml 目录下的 XML 文件,一般用来保存格式化的数据,在应用程序编译和打包时会将 XML 文件转换为高效的二进制格式,应用程序运行时会以特殊的方式进行访问。

ResourceFileDemo 示例演示了如何在程序运行时访问资源文件。当用户点击"读取原始文件"按钮时,程序将读取/res/raw/raw_file.txt 文件,并将内容显示在界面上,如图 11-11 所示。

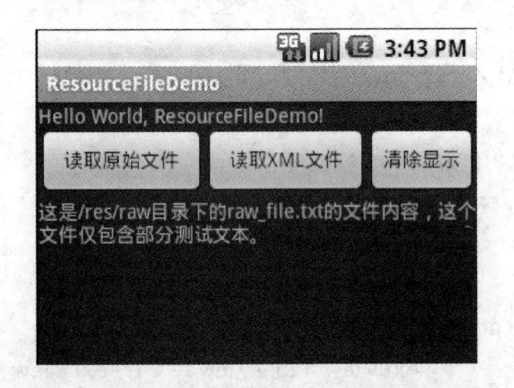

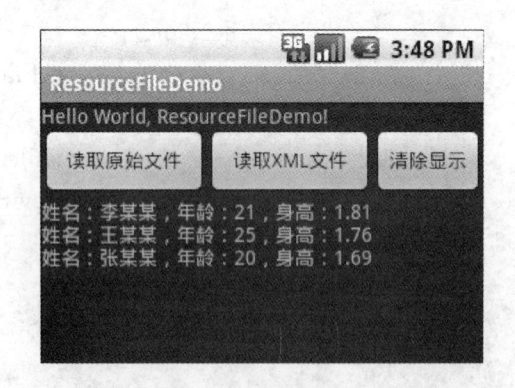

图 11-11 ResourceFileDemo 读取原始文件示例   图 11-12 ResourceFileDemo 读取 xml 文件示例

当用户点击"读取 XML 文件"按钮时,程序将读取/res/xml/people.xml 文件,并将内容显示在界面上,如图 11-12 所示。

读取原始格式文件,首先需要调用 getResource()函数获得资源对象,然后通过调用资源对象的 openRawResource()函数,以二进制流的形式打开指定的原始格式文件。在读取文件结束后,调用 close()函数关闭文件流。ResourceFileDemo 示例中关于读取原始格式文件的核心代码如下。

```
1. Resources resources = this.getResources();
2. InputStream inputStream = null;
3. try {
4.     inputStream = resources.openRawResource(R.raw.raw_file);
5.     byte[] reader = new byte[inputStream.available()];
6.     while (inputStream.read(reader) != -1) {
7.     }
8.     displayView.setText(new String(reader,"utf-8"));
9. } catch (IOException e) {
10.     Log.e("ResourceFileDemo", e.getMessage(), e);
11. } finally {
12.     if (inputStream != null) {
13.         try {
14.             inputStream.close();
15.         }
16.         catch (IOException e) { }
17.     }
18. }
```

代码第 8 行的 new String(reader,"utf-8"),表示以 UTF-8 的编码方式,从字节数组中实例化一个字符串。程序开发人员需要确定/res/raw/raw_file.txt 文件使用的是 UTF-8 编码方式,否则程序运行时会产生乱码。

确认文件的方法:右击 raw_file.txt 文件,选择"Properties"打开 raw_file.txt 文件的属性设置框,在 Resource 栏下的 Text file encoding 中,选择"Other:UTF-8"。如图 11-13 所示。

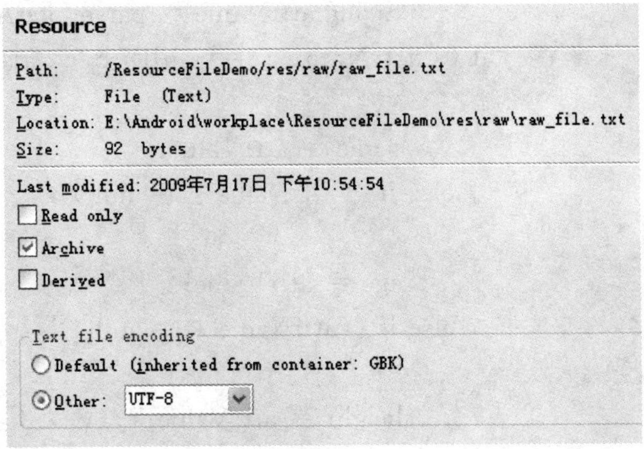

图 11-13 UTF-8 编码方式的确认

/res/xml 目录下的 XML 文件会转换为一种高效的二进制格式。

为了在程序运行时读取/res/xml 目录下的 XML 文件,首先在/res/xml 目录下创建一个名为 people.xml 的文件,XML 文件定义了多个<person>元素,每个<person>元素都包含三个属性 name、age 和 height,分别表示姓名、年龄和身高。/res/xml/people.xml 文件代码如下。

```
1. <people>
2.     <person name="李某某" age="21" height="1.81" />
3.     <person name="王某某" age="25" height="1.76" />
4.     <person name="张某某" age="20" height="1.69" />
5. </people>
```

接着读取 XML 格式文件,首先通过调用资源对象的 getXml()函数,获取到 XML 解析器 XmlPullParser。XmlPullParser 是 Android 平台标准的 XML 解析器,这项技术来自一个开源的 XML 解析 API 项目 XMLPULL。

ResourceFileDemo 示例中关于读取 XML 文件的核心代码如下。

```
1. XmlPullParser parser = resources.getXml(R.xml.people);
2. String msg = "";
3. try {
4.         while (parser.next() != XmlPullParser.END_DOCUMENT) {
5.             String people = parser.getName();
6.             String name = null;
7.                     String age = null;
8.             String height = null;
9.             if ((people != null) && people.equals("person")) {
10.                 int count = parser.getAttributeCount();
11.                 for (int i = 0; i < count; i++) {
12.                     String attrName = parser.getAttributeName(i);
13.                     String attrValue = parser.getAttributeValue(i);
14.                     if ((attrName != null) && attrName.equals("name")) {
15.                         name = attrValue;
16.                     } else if ((attrName != null) && attrName.equals("age")) {
17.                         age = attrValue;
18.                     } else if ((attrName != null) && attrName.equals("height")) {
19.                         height = attrValue;
20.                     }
21.                 }
```

第 11 章　数据存储和访问　　197

```
22.                    if ((name！＝null) && (age！＝null) &&
    (height！＝null)) {
23.                        msg＋＝"姓名："+name+"，年龄："+age+"，身高："
    +height+"\n";
24.                    }
25.                }
26.            }
27. } catch (Exception e) {
28.     Log.e("ResourceFileDemo", e.getMessage(), e);
29. }
30. displayView.setText(msg);
```

第 1 行代码通过资源对象的 getXml()函数获取到 XML 解析器。第 4 行代码的 parser. next()方法可以获取到高等级的解析事件，并通过对比确定事件类型。第 5 行代码使用 getName()函数获得元素的名称。第 10 行代码使用 getAttributeCount()函数获取元素的属性数量。第 12 行代码通过 getAttributeName()函数得到属性名称。第 14 行到第 19 行代码通过分析属性名获取到正确的属性值。第 23 行代码将属性值整理成需要显示的信息。

XmlPullParser 的 XML 事件类型见表 11-2 所示。

表 11-2　XmlPullParser 的 XML 事件类型

事件类型	说　　明
START_TAG	读取到标签开始标志
TEXT	读取文本内容
END_TAG	读取到标签结束标志
END_DOCUMENT	文档末尾

11.3　数据库存储

11.3.1　SQLite 数据库

SQLite 是一个开源的嵌入式关系数据库，在 2000 年由 D. Richard Hipp 发布。
SQLite 数据库特点：
① 更加适用于嵌入式系统，嵌入到使用它的应用程序中。
② 占用非常少，运行高效可靠，可移植性好。
③ 提供了零配置(zero-configuration)运行模式。

SQLite 数据库不仅提高了运行效率，而且屏蔽了数据库使用和管理的复杂性，程序仅需要进行最基本的数据操作，其他操作可以交给进程内部的数据库引擎完成。

SQLite 数据库采用了模块化设计，由 8 个独立的模块构成，这些独立模块又构成了三个主要的子系统，模块将复杂的查询过程分解为细小的工作进行处理，如图 11-14 所示。

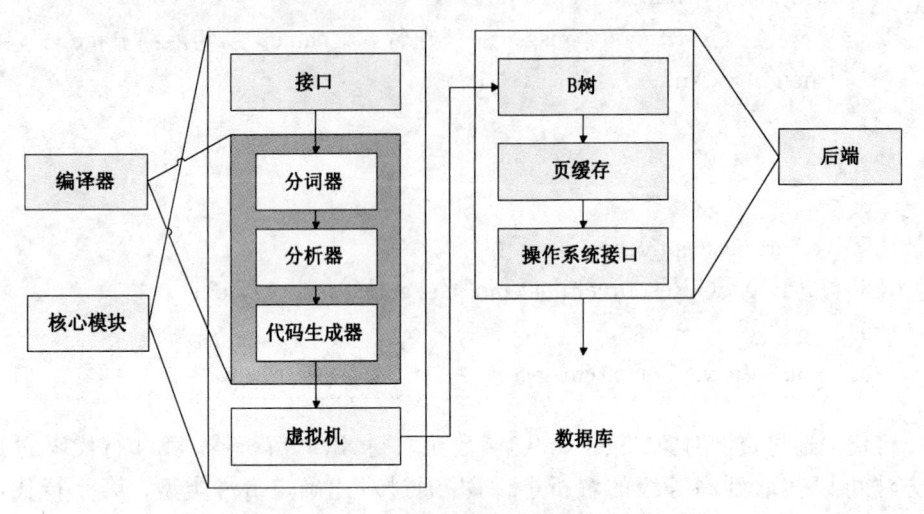

图 11-14　SQLite 数据库的模块化设计

接口由 SQLite C API 组成，因此无论是应用程序、脚本，还是库文件，最终都是通过接口与 SQLite 交互。

（1）编译器

由分词器和分析器组成。分词器和分析器对 SQL 语句进行语法检查，然后把 SQL 语句转化为底层能更方便处理的分层的数据结构，这种分层的数据结构称为"语法树"。把语法树传给代码生成器进行处理，生成一种针对 SQLite 的汇编代码，最后由虚拟机执行。

（2）虚拟机

SQLite 数据库体系结构中最核心的部分是虚拟机，也称为虚拟数据库引擎（Virtual Database Engine，VDBE）。与 Java 虚拟机相似，虚拟数据库引擎用来解释执行字节代码。虚拟数据库引擎的字节代码由 128 个操作码构成，这些操作码主要用以对数据库进行操作，每一条指令都可以完成特定的数据库操作，或以特定的方式处理栈的内容。

（3）后端

后端由 B-树、页缓存和操作系统接口构成。

B-树的主要功能就是索引，它维护着各个页面之间的复杂的关系，便于快速找到所需数据。页缓存的主要作用就是通过操作系统接口在 B-树和磁盘之间传递页面。B-树和页缓存共同对数据进行管理。

SQLite 数据库具有很强的移植性，可以运行在 Windows，Linux，BSD，Mac OS X 和一些商用 Unix 系统，比如 Sun 的 Solaris，IBM 的 AIX。SQLite 数据库也可以工作在许多嵌入式操作系统下，例如 QNX，VxWorks，Palm OS，Symbian 和 Windows CE。

SQLite 的核心大约有 3 万行标准 C 代码，模块化的设计使这些代码更加易于理解。

11.3.2　代码建库

在代码中动态建立数据库是比较常用的方法。

在程序运行过程中，当需要进行数据库操作时，应用程序会首先尝试打开数据库，此时如果数据库并不存在,程序则会自动建立数据库,然后再打开数据库。在编程实现时,一般将所有对数据库的操作都封装在一个类中,因此只要调用这个类,就可以完成对数据库的添加、更新、删除和查询等操作。

下面内容是 DBAdapter 类的部分代码,封装了数据库的建立、打开和关闭等操作。

```
1. public class DBAdapter {
2.     private static final String DB_NAME = "people.db";
3.     private static final String DB_TABLE = "peopleinfo";
4.     private static final int DB_VERSION = 1;
5.     public static final String KEY_ID = "_id";
6.     public static final String KEY_NAME = "name";
7.     public static final String KEY_AGE = "age";
8.     public static final String KEY_HEIGHT = "height";
9.     private SQLiteDatabase db;
10.    private final Context context;
11.    private DBOpenHelper dbOpenHelper;
12.
13. private static class DBOpenHelper extends SQLiteOpenHelper {}
14.
15.    public DBAdapter(Context _context) {
16.        context = _context;
17.    }
18.
19.    public void open() throws SQLiteException {
20.        dbOpenHelper = new DBOpenHelper(context, DB_NAME, null, DB_VERSION);
21.        try {
22.            db = dbOpenHelper.getWritableDatabase();
23.        }catch (SQLiteException ex) {
24.            db = dbOpenHelper.getReadableDatabase();
25.        }
26.    }
27.
28.        public void close() {
29.            if (db != null){
30.        db.close();
31.        db = null;
32.            }
33.    }
34. }
```

从代码的第2行到第9行可以看出，在DBAdapter类中首先声明了数据库的基本信息，包括数据库文件的名称、数据库表格名称和数据库版本，以及数据库表中的属性名称。从这些基本信息上不难发现，这个数据库与前一小节手动建立的数据库是完全相同的。

第11行代码声明了SQLiteDatabase对象db。SQLiteDatabase类封装了非常多的方法，用以建立、删除数据库，执行SQL命令，对数据进行管理等工作。第13行代码声明了一个非常重要的帮助类SQLiteOpenHelper，这个帮助类可以辅助建立、更新和打开数据库。第21行代码定义了open()函数用来打开数据库，但open()函数中并没有任何对数据库进行实际操作的代码，而是调用了SQLiteOpenHelper类的getWritableDatabase()函数和getReadableDatabase()函数。这两个函数会根据数据库是否存在、版本号和是否可写等情况，决定在返回数据库对象前，是否需要建立数据库。

在代码第30行的close()函数中，调用了SQLiteDatabase对象的close()方法关闭数据库，这是上面的代码中，唯一的一个地方直接调用了SQLiteDatabase对象的方法。

SQLiteDatabase中也封装了打开数据库的函数openDatabases()和创建数据库函数openOrCreateDatabases()，因为代码中使用了帮助类SQLiteOpenHelper，从而避免直接调用SQLiteDatabase中的打开和创建数据库的方法，简化了数据库打开过程中繁琐的逻辑判断过程。代码第15行实现了内部静态类DBOpenHelper，继承了帮助类SQLiteOpenHelper。

重载了onCreate()函数和onUpgrade()函数的代码如下：

```
1. private static class DBOpenHelper extends SQLiteOpenHelper {
2.     public DBOpenHelper(Context context, String name, CursorFactory factory, int version){
3.         super(context, name, factory, version);
4.     }
5.     private static final String DB_CREATE = "create table " +
6.         DB_TABLE + " ( " + KEY_ID + " integer primary key autoincrement, " +
7.             KEY_NAME + " text not null, " + KEY_AGE + " integer," + KEY_HEIGHT + " float);";
8.
9.     @Override
10.    public void onCreate(SQLiteDatabase _db) {
11.        db.execSQL(DB_CREATE);
12.    }
13.
14.    @Override
15.    public void onUpgrade(SQLiteDatabase _db, int _oldVersion, int _newVersion) {
16.        _db.execSQL("DROP TABLE IF EXISTS " + DB_TABLE);
```

```
17.            onCreate(_db);
18.        }
19. }
```

第 5 行到第 7 行代码的是创建表的 SQL 命令。第 10 行和第 15 行代码分别重载了 onCreate()函数和 onUpgrade()函数,这是继承 SQLiteOpenHelper 类必须重载的两个函数。onCreate()函数在数据库第一次建立时被调用,一般用来创建数据库中的表,并做适当的初始化工作。

在代码第 11 行中,通过调用 SQLiteDatabase 对象的 execSQL()方法,执行创建表的 SQL 命令。onUpgrade()函数在数据库需要升级时被调用,一般用来删除旧的数据库表,并将数据转移到新版本的数据库表中。第 16 行和第 17 行代码中,为了简单起见,并没有做任何的数据转移,而仅仅删除原有的表后建立新的数据库表。

程序开发人员不应直接调用 onCreate()和 onUpgrade()函数,而应该由 SQLiteOpenHelper 类来决定何时调用这两个函数。SQLiteOpenHelper 类的 getWritableDatabase()函数和 getReadableDatabase()函数是可以直接调用的函数。getWritableDatabase()函数用来建立或打开可读写的数据库对象,一旦函数调用成功,数据库对象将被缓存,任何需要使用数据库对象时,都可以调用这个方法获取到数据库对象,但一定要在不使用时调用 close()函数关闭数据库。如果保存数据库文件的磁盘空间已满,调用 getWritableDatabase()函数则无法获得可读写的数据库对象,这时可以调用 getReadableDatabase()函数,获得一个只读的数据库对象。

如果程序开发人员不希望使用 SQLiteOpenHelper 类,同样可以直接创建数据库。首先调用 openOrCreateDatabases()函数创建数据库对象,然后执行 SQL 命令建立数据库中的表和直接的关系,示例代码如下。

```
1. private static final String DB_CREATE = "create table " +
2.     DB_TABLE + " (" + KEY_ID + " integer primary key autoincrement, " +
3.     KEY_NAME+ " text not null, " + KEY_AGE+ " integer," + KEY
   _HEIGHT + " float);";
4. public void create() {
5.     db. openOrCreateDatabases(DB_NAME, context. MODE_PRIVATE, null)
6.     db. execSQL(DB_CREATE);
7. }
```

11.3.3 数据操作

数据操作是指对数据的添加、删除、查找和更新的操作。虽然可以通过执行 SQL 命名完成数据操作,但推荐使用 Android 提供的专用类和方法,这些类和方法更加简洁、易用。

为了使 DBAdapter 类支持对数据的添加、删除、更新和查找等功能,在 DBAdapter 类中增加下面的这些函数：
insert(People people)：用来添加一条数据。
queryAllData()：用来获取全部数据。
queryOneData(long id)：根据 id 获取一条数据。
deleteAllData()：用来删除全部数据。
deleteOneData(long id)：根据 id 删除一条数据。
updateOneData(long id，People people)：根据 id 更新一条数据。

```
1. public class DBAdapter {
2.      public long insert(People people) {}
3.      public long deleteAllData() { }
4.      public long deleteOneData(long id) { }
5.      public People[] queryAllData() {}
6.      public People[] queryOneData(long id) { }
7.      public long updateOneData(long id, People people) { }
8.
9.      private People[] ConvertToPeople(Cursor cursor){}
10. }
```

ConvertToPeople(Cursor cursor)是私有函数，作用是将查询结果转换为用来存储数据自定义的 People 类对象。People 类包含四个公共属性，分别为 ID、Name、Age 和 Height，对应数据库中的四个属性值。重载 toString()函数，主要是便于界面显示的需要。

People 类的代码如下。

```
1. public class People {
2.      public int ID = -1;
3.      public String Name;
4.      public int Age;
5.      public float Height;
6.
7.      @Override
8.      public String toString(){
9.          String result = "";
10.         result += "ID:" + this.ID + ",";
11.         result += "姓名:" + this.Name + ",";
12.         result += "年龄:" + this.Age + ",";
13.         result += "身高:" + this.Height + ",";
14.         return result;
15.     }
16. }
```

SQLiteDatabase 类的公共函数 insert()、delete()、update()和 query(),封装了执行的添加、删除、更新和查询功能的 SQL 命令。下面分别介绍如何使用 SQLiteDatabase 类的公共函数,完成数据的添加、删除、更新和查询等操作。

1. 添加功能

首先构造一个 ContentValues 对象,然后调用 ContentValues 对象的 put()方法,将每个属性的值写入到 ContentValues 对象中,最后使用 SQLiteDatabase 对象的 insert()函数,将 ContentValues 对象中的数据写入指定的数据库表中。

insert()函数的返回值是新数据插入的位置,即 ID 值。ContentValues 类是一个数据承载容器,主要用来向数据库表中添加一条数据。

```
1. public long insert(People people){
2.    ContentValues newValues = new ContentValues();
3.
4.    newValues.put(KEY_NAME, people.Name);
5.    newValues.put(KEY_AGE, people.Age);
6.    newValues.put(KEY_HEIGHT, people.Height);
7.
8.    return db.insert(DB_TABLE, null, newValues);
9. }
```

第 4 行代码向 ContentValues 对象 newValues 中添加一个名称/值对,put()函数的第 1 个参数是名称,第 2 个参数是值。在第 8 行代码的 insert()函数中,第 1 个参数是数据表的名称,第 2 个参数是在 NULL 时的替换数据,第 3 个参数是需要向数据库表中添加的数据。

2. 删除功能

删除数据比较简单,只需要调用当前数据库对象的 delete()函数,并指明表名称和删除条件即可。

```
1. public long deleteAllData(){
2.    return db.delete(DB_TABLE, null, null);
3. }
4. public long deleteOneData(long id){
5.    return db.delete(DB_TABLE, KEY_ID + "=" + id, null);
6. }
```

delete()函数的第 1 个参数是数据库的表名称,第 2 个参数是删除条件。在第 2 行代码中,删除条件为 null,表示删除表中的所有数据。第 6 行代码指明了需要删除数据的 id 值,因此 deleteOneData()函数仅删除一条数据,此时 delete()函数的返回值表示被删除的数据的数量。

3. 更新功能

更新数据同样要使用 ContentValues 对象,首先构造 ContentValues 对象,然后调用 put()函数将属性的值写入到 ContentValues 对象中,最后使用 SQLiteDatabase 对象的 update()

函数,并指定数据的更新条件。

```
1. public long updateOneData(long id, People people){
2.    ContentValues updateValues = new ContentValues();
3.    updateValues.put(KEY_NAME, people.Name);
4.    updateValues.put(KEY_AGE, people.Age);
5.    updateValues.put(KEY_HEIGHT, people.Height);
6.
7.    return db.update(DB_TABLE, updateValues, KEY_ID + "=" + id, null);
8. }
```

在代码的第 7 行中,update()函数的第 1 个参数表示数据表的名称,第 2 个参数是更新条件。update()函数的返回值表示数据库表中被更新的数据数量。

4. 查询功能

首先介绍 Cursor 类。在 Android 系统中,数据库查询结果的返回值并不是数据集合的完整拷贝,而是返回数据集的指针,这个指针就是 Cursor 类。Cursor 类支持在查询的数据集合中多种方式移动,并能够获取数据集合的属性名称和序号。

Cursor 类的方法和说明见表 11-3 所示。

表 11-3　Cursor 类的方法和说明

函数	说明
moveToFirst	将指针移动到第一条数据上
moveToNext	将指针移动到下一条数据上
moveToPrevious	将指针移动到上一条数据上
getCount	获取集合的数据数量
getColumnIndexOrThrow	返回指定属性名称的序号,如果属性不存在则产生异常
getColumnName	返回指定序号的属性名称
getColumnNames	返回属性名称的字符串数组
getColumnIndex	根据属性名称返回序号
moveToPosition	将指针移动到指定的数据上
getPosition	返回当前指针的位置

从 Cursor 中提取数据可以参考 ConvertToPeople()函数的实现方法。在提取 Cursor 数据中的数据前,推荐测试 Cursor 中的数据数量,避免在数据获取中产生异常,例如代码的第 3 行到第 5 行。使用类型安全的 get<Type>()函数从 Cursor 中提取数据,函数的输入值为属性的序号。为了获取属性的序号,可以使用 getColumnIndex()函数获取指定属性的序号。

```
1. private People[] ConvertToPeople(Cursor cursor){
2.     int resultCounts = cursor.getCount();
3.     if (resultCounts == 0 || ! cursor.moveToFirst()){
4.         return null;
5.     }
6.     People[] peoples = new People[resultCounts];
7.     for (int i = 0 ; i<resultCounts; i++){
8.         peoples[i] = new People();
9.         peoples[i].ID = cursor.getInt(0);
10.        peoples[i].Name = cursor.getString(cursor.getColumnIndex(KEY_NAME));
11.        peoples[i].Age = cursor.getInt(cursor.getColumnIndex(KEY_AGE));
12.        peoples[i].Height = cursor.getFloat(cursor.getColumnIndex(KEY_HEIGHT));
13.        cursor.moveToNext();
14.    }
15.    return peoples;
16. }
```

要进行数据查询就需要调用 SQLiteDatabase 类的 query() 函数, query() 函数的语法如下。

> Cursor android.database.sqlite.SQLiteDatabase.query(String table, String[] columns, String selection, String[] selectionArgs, String groupBy, String having, String orderBy)

query() 函数的参数说明如表 11-4 所示。

表 11-4 query() 函数的参数说明

位置	类型+名称	说明
1	String table	表名称
2	String[] columns	返回的属性列名称
3	String selection	查询条件
4	String[] selectionArgs	如果在查询条件中使用的问号,则需要定义替换符的具体内容
5	String groupBy	分组方式
6	String having	定义组的过滤器
7	String orderBy	排序方式

根据 id 查询数据的代码如下。

```
1. public People[] getOneData(long id) {
2.     Cursor results =   db.query(DB_TABLE, new String[] { KEY_ID,
   KEY_NAME, KEY_AGE, KEY_HEIGHT}, KEY_ID + "=" + id, null,
   null, null, null);
3.     return ConvertToPeople(results);
4. }
```

根据 id 查询全部数据的代码如下。

```
1. public People[] getAllData() {
2.     Cursor results = db.query(DB_TABLE, new String[] { KEY_ID, KEY_
   NAME, KEY_AGE, KEY_HEIGHT}, null, null, null, null, null);
3.     return ConvertToPeople(results);
4. }
```

SQLiteDemo 是对数据库操作的一个示例，如图 11-15 所示。

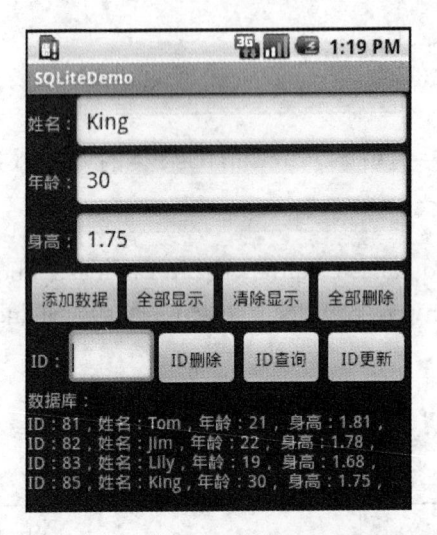

图 11-15　SQLiteDemo 的数据库操作示例

用户可以在界面的上方输入数据信息，通过"添加数据"按钮将数据写入数据库。"全部显示"相当于查询数据库中的所有数据，并将数据显示在界面下方；"清除显示"仅是清除界面下面显示的数据，而不对数据库进行任何操作；"全部删除"是数据库操作，将删除数据库中的所有数据。

在界面中部，以"ID+功能"命名的按钮，分别是根据 ID 删除数据，根据 ID 进行数据查询和更新，而这个 ID 值就取自本行的 EditText 控件。

11.4 数据分享

11.4.1 ContentProvider

ContentProvider(数据提供者)是在应用程序间共享数据的一种接口机制,它提供了更为高级的数据共享方法,应用程序可以指定需要共享的数据,而其他应用程序则可以在不知数据来源、路径的情况下,对共享数据进行查询、添加、删除和更新等操作。许多 Android 系统的内置数据也通过 ContentProvider 提供给用户使用,例如通讯录、音视频文件和图像文件等。

在创建 ContentProvider 时,需要首先使用数据库、文件系统或网络实现底层存储功能,然后在继承 ContentProvider 的类中实现基本数据操作的接口函数,包括添加、删除、查找和更新等功能。调用者不能够直接调用 ContentProvider 的接口函数,而需要使用 ContentResolver 对象,通过 URI 间接调用 ContentProvider。下图 11 - 16 是 ContentProvider 的调用关系。

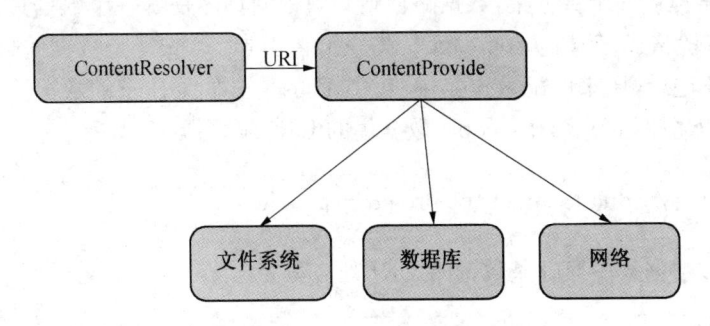

图 11 - 16 ContentProvider 调用关系

程序开发人员使用 ContentResolver 对象与 ContentProvider 进行交互,而 ContentResolver 则通过 URI 确定需要访问的 ContentProvider 的数据集。在发起一个请求的过程中,Android 首先根据 URI 确定处理这个查询的 ContentResolver,然后初始化 ContentResolver 所有需要的资源,这个初始化的工作是 Android 系统完成的,无需程序开发人员参与。一般情况下只有一个 ContentResolver 对象,但却可以同时与多个 ContentProvider 进行交互。

ContentProvider 完全屏蔽了数据提供组件的数据存储方法。在使用者看来,数据提供者通过 ContentProvider 提供了一组标准的数据操作接口,却无法得知数据提供者的数据存储方。数据提供者可以使用 SQLite 数据库存储数据,也可以通过文件系统或 SharedPreferences 存储数据,甚至是使用网络存储的方法,这些内容对数据使用者都是不可见。同时也正是因为屏蔽数据的存储方法,很大程度上简化的 ContentProvider 的使用难度,使用者只要调用 ContentProvider 提供的接口函数,就可完成所有的数据操作。

ContentProvider 的数据模式似于数据库的数据表,每行是一条记录,每列具有相同的数据类型。每条记录都包含一个长型的字段_ID,用来唯一标识每条记录。ContentProvider 可以提供多个数据集,调用者使用 URI 对不同的数据集的数据进行操作。ContentProvider 数据模型,如表 11 - 5 所示。

表 11-5 ContentProvider 数据模型

_ID	NAME	AGE	HEIGHT
1	Tom	21	1.81
2	Jim	22	1.78

URI 是通用资源标志符(Uniform Resource Identifier),用来定位任何远程或本地的可用资源。ContentProvider 使用的 URI 语法结构如下。

content://<authority>/<data_path>/<id>

content:// 是通用前缀,表示该 URI 用于 ContentProvider 定位资源,无需修改。<authority>是授权者名称,用来确定具体由哪一个 ContentProvider 提供资源。因此,一般<authority>都由类的小写全称组成,以保证唯一性。<data_path>是数据路径,用来确定请求的是哪个数据集。

如果 ContentProvider 仅提供一个数据集,数据路径则是可以省略的;如果 ContentProvider 提供多个数据集,数据路径则必须指明具体是哪一个数据集。数据集的数据路径可以写成多段格式,例如/people/girl 和/people/boy。<id>是数据编号,用来唯一确定数据集中的一条记录,用来匹配数据集中_ID 字段的值。如果请求的数据并不只限于一条数据,则<id>可以省略,请求整个 people 数据集的 URI 应写为:

content://edu.xzceu.peopleprovider/people

请求 people 数据集中第 3 条数据的 URI 则应写为:

content://edu.xzceu.peopleprovider/people/3

11.4.2 创建数据提供者

程序开发人员通过继承 ContentProvider 类可以创建一个新的数据提供者,过程可以分为三步:

① 继承 ContentProvider,并重载六个函数。
② 声明 CONTENT_URI,实现 UriMatcher。
③ 注册 ContentProvider。

继承 ContentProvider 后重载的六个函数分别如下:
① delete():删除数据集。
② insert():添加数据集。
③ qurey():查询数据集。
④ update():更新数据集。
⑤ onCreate():初始化底层数据集和建立数据连接等工作。
⑥ getType():返回指定 URI 的 MIME 数据类型。
注意:如果 URI 是单条数据,则返回的 MIME 数据类型应以 vnd.android.cursor.item 开

头；如果 URI 是多条数据，则返回的 MIME 数据类型应以 vnd. android. cursor. dir/ 开头。

新建立的类继承 ContentProvider 后，Eclipse 会提示程序开发人员需要重载部分代码，并自动生成需要重载的代码框架。下面的代码是 Eclipse 自动生成的代码框架。

```
1. import android. content. * ;
2. import android. database. Cursor;
3. import android. net. Uri;
4. public class PeopleProvider extends ContentProvider{
5.      @Override
6.      public int delete(Uri uri, String selection, String[] selectionArgs) {
7.          // TODO Auto-generated method stub
8.          return 0;
9.      }
10.
11.     @Override
12.     public String getType(Uri uri) {
13.         // TODO Auto-generated method stub
14.         return null;
15.     }
16.
17.     @Override
18.     public Uri insert(Uri uri, ContentValues values) {
19.         // TODO Auto-generated method stub
20.         return null;
21.     }
22.
23.     @Override
24.     public boolean onCreate() {
25.         // TODO Auto-generated method stub
26.         return false;
27.     }
28.
29.     @Override
30.     public Cursor query(Uri uri, String[] projection, String selection,
31.         String[] selectionArgs, String sortOrder) {
32.         // TODO Auto-generated method stub
33.         return null;
34.     }
35.
36.     @Override
```

```
37.     public int update(Uri uri, ContentValues values, String selection,
38.             String[] selectionArgs) {
39.         // TODO Auto-generated method stub
40.         return 0;
41.     }
42. }
```

1. 声明 CONTENT_URI,实现 UriMatcher

在新构造的 ContentProvider 类中,通过构造一个 UriMatcher,判断 URI 是单条数据还是多条数据。为了便于判断和使用 URI,一般将 URI 的授权者名称和数据路径等内容声明为静态常量,并声明 CONTENT_URI。声明 CONTENT_URI 和构造 UriMatcher 的代码如下。

```
1. public static final String AUTHORITY = "edu.xzceu.peopleprovider";
2. public static final String PATH_SINGLE = "people/#";
3. public static final String PATH_MULTIPLE = "people";
4. public static final String CONTENT_URI_STRING = "content://" +
    AUTHORITY + "/" + PATH_MULTIPLE;
5. public static final Uri CONTENT_URI = Uri.parse(CONTENT_URI_STRING);
6. private static final int MULTIPLE_PEOPLE = 1;
7. private static final int SINGLE_PEOPLE = 2;
8.
9. private static final UriMatcher uriMatcher;
10. static {
11.     uriMatcher = new UriMatcher(UriMatcher.NO_MATCH);
12.     uriMatcher.addURI(AUTHORITY, PATH_SINGLE, MULTIPLE_PEOPLE);
13.     uriMatcher.addURI(AUTHORITY, PATH_MULTIPLE, SINGLE_PEOPLE);
14. }
```

第 1 行代码声明了 URI 的授权者名称。第 2 行代码声明了单条数据的数据路径。第 3 行代码声明了多条数据的数据路径。第 4 行代码声明了 CONTENT_URI 的字符串形式。第 5 行代码则正式声明了 CONTENT_URI。第 6 行代码声明了多条数据的返回代码。第 7 行代码声明了单条数据的返回代码。第 9 行代码声明了 UriMatcher。第 10 行到第 13 行的静态构造函数中,声明了 UriMatcher 的匹配方式和返回代码。其中第 11 行 UriMatcher 的构造函数中,UriMatcher.NO_MATCH 表示 URI 无匹配时的返回代码。第 12 行的 addURI()函数用来添加新的匹配项,语法如下。

```
public void addURI (String authority, String path, int code)
```

authority 表示匹配的授权者名称,path 表示数据路径,♯可以代表任何数字,code 表示返回代码。

声明 CONTENT_URI,实现 UriMatcher。使用 UriMatcher 时,则可以直接调用 match()函数,对指定的 URI 进行判断,示例代码如下。

```
1. switch(uriMatcher.match(uri)){
2.     case MULTIPLE_PEOPLE:
3.         //多条数据的处理过程
4.         break;
5.     case SINGLE_PEOPLE:
6.         //单条数据的处理过程
7.         break;
8.     default:
9.         throw new IllegalArgumentException("不支持的 URI:" + uri);
10. }
```

2. 注册 ContentProvider

在完成 ContentProvider 类的代码实现后,需要在 AndroidManifest.xml 文件中进行注册。注册 ContentProvider 使用<provider>标签,示例代码如下。

```
1. <application android:icon="@drawable/icon"
   android:label="@string/app_name">
2.     <provider android:name = ".PeopleProvider"
3.         android:authorities = "edu.xzceu.peopleprovider"/>
4. </application>
```

在上面的代码中,注册了一个授权者名称为 edu.xzceu.peopleprovider 的 ContentProvider,其实现类是 PeopleProvider。

使用 ContentProvider 是通过 Android 组件都具有的 ContentResolver 对象,通过 URI 进行数据操作。程序开发人员只需要知道 URI 和数据集的数据格式,则可以进行数据操作,解决不同应用程序之间的数据共享问题。每个 Android 组件都具有一个 ContentResolver 对象,获取 ContentResolver 对象的方法是调用 getContentResolver()函数。

```
ContentResolver resolver = getContentResolver();
```

3. 查询操作

在获取到 ContentResolver 对象后,程序开发人员则可以使用 query()函数查询目标数据。下面的代码是查询 ID 为 2 的数据,其中在 URI 中定义了需要查询数据的 ID,在 query()函数并没有额外声明查询条件。

```
1. String KEY_ID = "_id";
2. String KEY_NAME = "name";
```

```
3. String KEY_AGE = "age";
4. String KEY_HEIGHT = "height";
5.
6. Uri uri = Uri.parse(CONTENT_URI_STRING + "/" + "2";
7. Cursor cursor = resolver.query(uri,
8.      new String[]{KEY_ID, KEY_NAME, KEY_AGE, KEY_HEIGHT}, null,
   null, null);
```

如果需要获取数据集中的全部数据,则可直接使用 CONTENT_URI,此时 ContentProvider 在分析 URI 时将认为需要返回全部数据。

ContentResolver 的 query()函数与 SQLite 数据库的 query()函数非常相似,语法结构如下。

```
Cursor query(Uri uri, String[] projection, String selection, String[] selectionArgs, String sortOrder)
```

uri 定义了查询的数据集,projection 定义了从数据集返回哪些数据项,selection 定义了返回数据的查询条件。

4. 添加操作

向 ContentProvider 中添加数据有两种方法:一种是使用 insert()函数,向 ContentProvider 中添加一条数据;另一种是使用 bultInsert()函数,批量的添加数据。下面的代码说明了如何使用 insert()函数添加单条数据。

```
1. ContentValues values = new ContentValues();
2. values.put(KEY_NAME, "Tom");
3. values.put(KEY_AGE, 21);
4. values.put(KEY_HEIGHT, );
5. Uri newUri = resolver.insert(CONTENT_URI, values);
```

下面代码说明了如何使用 bultInsert()函数添加多条数据。

```
1. ContentValues[] arrayValues = new ContentValues[10];
2. //实例化每一个 ContentValues
3. int count = resolver.bultInsert(CONTENT_URI, arrayValues);
```

5. 删除操作

删除操作需要使用 delete()函数。如果需要删除单条数据,则可以在 URI 中指定需要删除数据的 ID;如果需要删除多条数据,则可以在 selection 中声明删除条件。下面代码说明了如何删除 ID 为 2 的数据。

```
1. Uri uri = Uri.parse(CONTENT_URI_STRING + "/" + "2");
2. int result = resolver.delete(uri, null, null);
```

也可以在 selection 将删除条件定义为 ID 大于 4 的数据。

> 1. String selection = KEY_ID + ">4";
> 2. int result = resolver.delete(CONTENT_URI, selection, null);

6. 更新操作

更新操作需要使用 update() 函数，参数定义与 delete() 函数相同，同样可以在 URI 中指定需要更新数据的 ID，也可以在 selection 中声明更新条件。下面代码说明了如何更新 ID 为 7 的数据。

> 1. ContentValues values = new ContentValues();
> 2. values.put(KEY_NAME, "Tom");
> 3. values.put(KEY_AGE, 21);
> 4. values.put(KEY_HEIGHT,);
> 5. Uri uri = Uri.parse(CONTENT_URI_STRING + "/" + "7");
> 6. int result = resolver.update(uri, values, null, null);

11.4.3 数据分享示例

ContentProviderDemo 是一个无界面的示例，仅提供一个 ContentProvider 组件，供其他应用程序进行数据交换。底层使用 SQLite 数据库，支持数据的添加、删除、更新和查询等基本操作。ContentResolverDemo 是使用 ContentProvider 的示例，自身不具有任何数据存储功能，仅是通过 URI 访问 ContentProviderDemo 示例提供的 ContentProvider。如图 11-17 所示。

图 11-17 ContentResolverDemo 的示例　　　图 11-18 ContentProviderDemo 文件结构

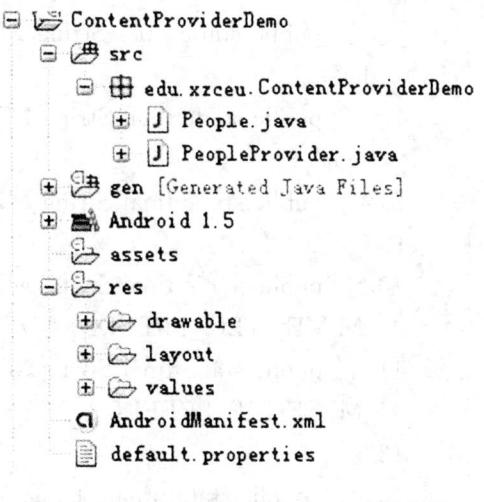

ContentProviderDemo 文件结构如图 11-18 所示。
ContentResolverDemo 文件结构如图 11-19 所示。

```
ContentResolverDemo
  src
    edu.xzceu.ContentResolverDemo
      ContentResolverDemo.java
      People.java
  gen [Generated Java Files]
  Android 1.5
  assets
  res
    drawable
    layout
      main.xml
    values
  AndroidManifest.xml
  default.properties
```

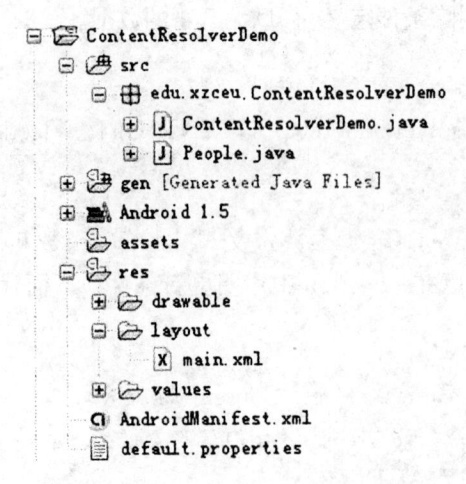

图 11-19 ContentResolverDemo 文件结构

以上两个示例都包含一个相同的文件 People.java，两个示例中的这个文件的内容也完全相同，定义了数据提供者和数据调用者都必须知道的信息。这些信息包括授权者名称、数据路径、MIME 数据类型、CONTENT_URI 和数据项名称等。下面分别给出 People.java、PeopleProvider.java 和 ContentResolverDemo.java 的完整代码，最后分别给出 ContentProviderDemo 示例和 ContentResolverDemo 示例的 AndroidManifest.xml 文件内容。

People.java 文件的完整代码如下：

```
1. package edu.xzceu.ContentResolverDemo;
2. import android.net.Uri;
3.
4. public class People{
5.
6.     public static final String MIME_DIR_PREFIX = "vnd.android.cursor.dir";
7.     public static final String MIME_ITEM_PREFIX = "vnd.android.cursor.item";
8.     public static final String MINE_ITEM = "vnd.xzceu.people";
9.
10.    public static final String MINE_TYPE_SINGLE = 
       MIME_ITEM_PREFIX + "/" + MINE_ITEM;
11.    public static final String MINE_TYPE_MULTIPLE = 
       MIME_DIR_PREFIX + "/" + MINE_ITEM;
12.
13.    public static final String AUTHORITY = "edu.xzceu.peopleprovider";
14.    public static final String PATH_SINGLE = "people/#";
15.    public static final String PATH_MULTIPLE = "people";
```

```
16.     public static final String CONTENT_URI_STRING = "content://" +
   AUTHORITY + "/" + PATH_MULTIPLE;
17.     public static final Uri  CONTENT_URI = Uri.parse(CONTENT_URI
   _STRING);
18.     public static final String KEY_ID = "_id";
19.     public static final String KEY_NAME = "name";
20.     public static final String KEY_AGE = "age";
21.     public static final String KEY_HEIGHT = "height";
22. }
```

PeopleProvider.java 文件的完整代码如下。

```
1. package edu.xzceu.ContentProviderDemo;
2. import android.content.ContentProvider;
3. import android.content.ContentUris;
4. import android.content.ContentValues;
5. import android.content.Context;
6. import android.content.UriMatcher;
7. import android.database.Cursor;
8. import android.database.SQLException;
9. import android.database.sqlite.SQLiteDatabase;
10. import android.database.sqlite.SQLiteOpenHelper;
11. import android.database.sqlite.SQLiteQueryBuilder;
12. import android.database.sqlite.SQLiteDatabase.CursorFactory;
13. import android.net.Uri;
14. public class PeopleProvider extends ContentProvider{
15.     private static final String DB_NAME = "people.db";
16.     private static final String DB_TABLE = "peopleinfo";
17.     private static final int DB_VERSION = 1;
18.     private SQLiteDatabase db;
19.     private DBOpenHelper dbOpenHelper;
20.     private static final int MULTIPLE_PEOPLE = 1;
21.     private static final int SINGLE_PEOPLE = 2;
22.     private static final UriMatcher uriMatcher;
23.         static {
24.         uriMatcher = new UriMatcher(UriMatcher.NO_MATCH);
25.         uriMatcher.addURI(People.AUTHORITY, People.PATH_
   MULTIPLE, MULTIPLE_PEOPLE);
```

```
26.            uriMatcher.addURI(People.AUTHORITY, People.PATH_
   SINGLE, SINGLE_PEOPLE);
27.        }
28.
29.        @Override
30.        public String getType(Uri uri) {
31.                    switch(uriMatcher.match(uri)){
32.                case MULTIPLE_PEOPLE:
33.                    return People.MINE_TYPE_MULTIPLE;
34.                case SINGLE_PEOPLE:
35.                    return People.MINE_TYPE_SINGLE;
36.                default:
37.                    throw new
   IllegalArgumentException("Unkown uri:"+uri);
38.            }
39.            }
40.
41.        @Override
42.        public int delete(Uri uri, String selection, String[] selectionArgs) {
43.                int count = 0;
44.                switch(uriMatcher.match(uri)){
45.                case MULTIPLE_PEOPLE:
46.                    count = db.delete(DB_TABLE, selection, selectionArgs);
47.                    break;
48.                    case SINGLE_PEOPLE:
49.                    String segment = uri.getPathSegments().get(1);
50.                        count = db.delete(DB_TABLE, People.KEY_ID
   + "=" + segment, selectionArgs);
51.                    break;
52.                default:
53.                    throw new
   IllegalArgumentException("Unsupported URI:" + uri);
54.            }
55. getContext().getContentResolver().notifyChange(uri, null);
56.        return count;
57.        }
58.
59.        @Override
60.        public Uri insert(Uri uri, ContentValues values) {
```

```
61.            long id = db.insert(DB_TABLE, null, values);
62.            if ( id > 0 ){
63.               Uri newUri ContentUris.withAppendedId(People.CONTENT_URI, id);
64.               getContext().getContentResolver().notifyChange(newUri, null);
65.               return newUri;
66.            }
67.            throw new SQLException("Failed to insert row into " + uri);
68.         }
69.
70.         @Override
71.         public boolean onCreate() {
72.            Context context = getContext();
73.            dbOpenHelper = new DBOpenHelper(context, DB_NAME, null, DB_VERSION);
74.            db = dbOpenHelper.getWritableDatabase();
75.
76.            if (db == null)
77.               return false;
78.            else
79.               return true;
80.         }
81.
82.         @Override
83.         public Cursor query(Uri uri, String[] projection, String selection,
84.            String[] selectionArgs, String sortOrder) {
85.            SQLiteQueryBuilder qb = new SQLiteQueryBuilder();
86.            qb.setTables(DB_TABLE);
87.            switch(uriMatcher.match(uri)){
88.               case SINGLE_PEOPLE:
89.                  qb.appendWhere(People.KEY_ID + "=" + uri.getPathSegments().get(1));
90.                  break;
91.               default:
92.                  break;
93.            }
94.            Cursor cursor = qb.query(db,
95.               projection,
```

```
96.            selection,
97.            selectionArgs,
98.            null,
99.            null,
100.           sortOrder);
101.       cursor.setNotificationUri(getContext().getContentResolver(), uri);
102.       return cursor;
103.     }
104.
105.     @Override
106.     public int update(Uri uri, ContentValues values, String selection,
107.         String[] selectionArgs) {
108.       int count;
109.       switch(uriMatcher.match(uri)){
110.         case MULTIPLE_PEOPLE:
111.             count = db.update(DB_TABLE, values, selection, selectionArgs);
112.             break;
113.         case SINGLE_PEOPLE:
114.             String segment = uri.getPathSegments().get(1);
115.             count = db.update(DB_TABLE, values, People.KEY_ID+"="+segment, selectionArgs);
116.             break;
117.         default:
118.             throw new IllegalArgumentException("Unknow URI:" + uri);
119.       }
120.       getContext().getContentResolver().notifyChange(uri, null);
121.       return count;
122.     }
123.
124.   private static class DBOpenHelper extends SQLiteOpenHelper {
125.     public DBOpenHelper(Context context, String name, CursorFactory factory, int version) {
126.         super(context, name, factory, version);
127.     }
128. }
```

ContentResolverDemo.java 文件的完整代码如下。

```java
1. package edu.xzceu.ContentResolverDemo;
2. import android.app.Activity;
3. import android.content.ContentResolver;
4. import android.content.ContentValues;
5. import android.database.Cursor;
6. import android.net.Uri;
7. import android.os.Bundle;
8. import android.view.View;
9. import android.view.View.OnClickListener;
10. import android.widget.Button;
11. import android.widget.EditText;
12. import android.widget.TextView;
13. public class ContentResolverDemo extends Activity {
14.     private EditText nameText;
15.     private EditText ageText;
16.     private EditText heightText;
17.     private EditText idEntry;
18.     private TextView labelView;
19.     private TextView displayView;
20.     private ContentResolver resolver;
21.
22.     @Override
23.     public void onCreate(Bundle savedInstanceState) {
24.         super.onCreate(savedInstanceState);
25.         setContentView(R.layout.main);
26.         nameText = (EditText)findViewById(R.id.name);
27.         ageText = (EditText)findViewById(R.id.age);
28.         heightText = (EditText)findViewById(R.id.height);
29.         idEntry = (EditText)findViewById(R.id.id_entry);
30.         labelView = (TextView)findViewById(R.id.label);
31.         displayView = (TextView)findViewById(R.id.display);
32.         Button addButton=(Button)findViewById(R.id.add);
33.         Button queryAllButton=(Button)findViewById(R.id.query_all);
34.         Button clearButton=(Button)findViewById(R.id.clear);
35.         Button deleteAllButton=(Button)findViewById(R.id.delete_all);
36.         Button queryButton=(Button)findViewById(R.id.query);
```

```
37.         Button deleteButton=
   (Button)findViewById(R.id.delete);
38.         Button updateButton=
   (Button)findViewById(R.id.update);
39.         addButton.setOnClickListener(addButtonListener);
40.         queryAllButton.setOnClickListener(queryAllButtonListener);
41.         clearButton.setOnClickListener(clearButtonListener);
   deleteAllButton.setOnClickListener(deleteAllButtonListener);
42.         queryButton.setOnClickListener(queryButtonListener);
43.         deleteButton.setOnClickListener(deleteButtonListener);
44.         updateButton.setOnClickListener(updateButtonListener);
45.         resolver = this.getContentResolver();
46.     }
47.
48.     OnClickListener addButtonListener = new OnClickListener() {
49.         @Override
50.         public void onClick(View v) {
51.             ContentValues values = new ContentValues();
52.             values.put(People.KEY_NAME, nameText.getText().toString());
53.             values.put(People.KEY_AGE, Integer.parseInt(ageText.getText().toString()));
54.             values.put(People.KEY_HEIGHT, Float.parseFloat(heightText.getText().toString()));
55.             Uri newUri = resolver.insert(People.CONTENT_URI, values);
56.             labelView.setText("添加成功,URI:" + newUri);
57.         }
58.     };
59.
60.     OnClickListener queryAllButtonListener = new OnClickListener() {
61.         @Override
62.         public void onClick(View v) {
63.             Cursor cursor = resolver.query(People.CONTENT_URI,\
64.                 new String[]{ People.KEY_ID, People.KEY_NAME, People.KEY_AGE, People.KEY_HEIGHT},
65.                 null, null, null);
66.             if (cursor == null){
```

```java
67.                    labelView.setText("数据库中没有数据");
68.                    return;
69.                }
70.                labelView.setText("数据库:" + String.valueOf(cursor.getCount()) + "条记录");
71.                String msg = "";
72.                if(cursor.moveToFirst()){
73.                    do{
74.                        msg += " ID:" + cursor.getInt(cursor.getColumnIndex(People.KEY_ID)) + ",";
75.                        msg += "姓名:" + cursor.getString(cursor.getColumnIndex(People.KEY_NAME)) + ",";
76.                        msg += "年龄:" + cursor.getInt(cursor.getColumnIndex(People.KEY_AGE)) + ", ";
77.                        msg += "身高:" + cursor.getFloat(cursor.getColumnIndex(People.KEY_HEIGHT)) + "\n";
78.                    }while(cursor.moveToNext());
79.                }
80.                displayView.setText(msg);
81.            }
82.        };
83.
84.        OnClickListener clearButtonListener = new OnClickListener() {
85.
86.            @Override
87.            public void onClick(View v) {
88.                displayView.setText("");
89.            }
90.        };
91.        OnClickListener deleteAllButtonListener = new OnClickListener() {
92.            @Override
93.            public void onClick(View v) {
94.                resolver.delete(People.CONTENT_URI, null, null);
95.                String msg = "数据全部删除";
96.                labelView.setText(msg);
97.            }
98.        };
99.        };
```

```java
100.
101.    OnClickListener queryButtonListener = new OnClickListener() {
102.        @Override
103.        public void onClick(View v) {
104.            Uri uri = Uri.parse(People.CONTENT_URI_STRING + "/" + idEntry.getText().toString());
105.            Cursor cursor = resolver.query(uri,
106.                    new String[] { People.KEY_ID, People.KEY_NAME, People.KEY_AGE, People.KEY_HEIGHT},
107.                    null, null, null);
108.            if (cursor == null){
109.                labelView.setText("数据库中没有数据");
110.                return;
111.            }
112.
113.            String msg = "";
114.            if (cursor.moveToFirst()){
115.                msg += "ID:" + cursor.getInt(cursor.getColumnIndex(People.KEY_ID)) + ",";
116.                msg += "姓名:" + cursor.getString(cursor.getColumnIndex(People.KEY_NAME))+ ",";
117.                msg += "年龄:" + cursor.getInt(cursor.getColumnIndex(People.KEY_AGE)) + ",";
118.                msg += "身高:" + cursor.getFloat(cursor.getColumnIndex(People.KEY_HEIGHT)) + "\n";
119.            }
120.
121.            labelView.setText("数据库:");
122.            displayView.setText(msg);
123.        }
124.    };
125.
126.    OnClickListener deleteButtonListener= new OnClickListener() {
127.        @Override
128.        public void onClick(View v) {
129.            Uri uri = Uri.parse(People.CONTENT_URI_STRING + "/" + idEntry.getText().toString());
```

```
130.                int result = resolver.delete(uri, null, null);
131.                String msg = "删除ID为"+idEntry.getText().toString()
    +"的数据" + (result>0?"成功":"失败");
132.                labelView.setText(msg);
133.            }
134.        };
135.
136.        OnClickListener updateButtonListener = new OnClickListener() {
137.            @Override
138.            public void onClick(View v) {
139.                ContentValues values = new ContentValues();
140.
141. values.put(People.KEY_NAME, nameText.getText().toString());
142.                values.put(People.KEY_AGE, Integer.parseInt(ageText.
    getText().toString()));
143.                values.put(People.KEY_HEIGHT,
    Float.parseFloat(heightText.getText().toString()));
144.                Uri uri = Uri.parse(People.CONTENT_URI_STRING +
    "/" + idEntry.getText().toString());
145.                int result = resolver.update(uri, values, null, null);
146.                String msg = "更新ID为"+idEntry.getText().toString()
    +"的数据" + (result>0?"成功":"失败");
147.                labelView.setText(msg);
148.            }
149.        };
150. }
```

ContentProviderDemo 示例的 AndroidManifest.xml 文件内容如下。

```
1. <?xml version="1.0" encoding="utf-8"?>
2. <manifest xmlns:android="http://schemas.android.com/apk/res/android"
3.        package="edu.xzceu.ContentProviderDemo"
4.        android:versionCode="1"
5.        android:versionName="1.0">
6.     <application android:icon="@drawable/icon" android:label="@string/app_name">
7.        <provider android:name = ".PeopleProvider"
8.            android:authorities = "edu.xzceu.peopleprovider"/>
9.     </application>
10.    <uses-sdk android:minSdkVersion="3" />
11. </manifest>
```

ContentResolverDemo 示例的 AndroidManifest.xml 文件内容如下。

```
1. <? xml version="1.0" encoding="utf-8"? >
2. <manifest
   xmlns:android="http://schemas.android.com/apk/res/android"
3.        package="edu.xzceu.ContentResolverDemo"
4.        android:versionCode="1"
5.        android:versionName="1.0">
6.    <application android:icon="@drawable/icon"
      android:label="@string/app_name">
7.        <activity android:name=".ContentResolverDemo"
8.                  android:label="@string/app_name">
9.            <intent-filter>
10.               <action
      android:name="android.intent.action.MAIN" />
11.               <category
      android:name="android.intent.category.LAUNCHER" />
12.           </intent-filter>
13.       </activity>
14.   </application>
15.   <uses-sdk android:minSdkVersion="3" />
16. </manifest>
```

习题与思考题

1. 应用程序一般允许用户自己定义配置信息，如界面背景颜色、字体大小和字体颜色等，尝试使用 SharedPreferences 保存用户的自定义配置信息，并在程序启动时自动加载这些自定义的配置信息。

2. 尝试把第 1 题的用户自己定义配置信息，以 INI 文件的形式保存在内部存储器上。

3. 简述在嵌入式系统中使用 SQLite 数据库的优势。

4. 分别使用手动建库和代码建库的方式，创建名为 test.db 的数据库，并建立 staff 数据表，表内的属性值如下表 11-6 所示。

表 11-6 staff 数据表属性值

属　性	数据类型	说　明
_id	integer	主键
name	text	姓名
sex	text	性别
department	text	所在部门
salary	float	工资

5. 利用第 4 题所建立的数据库和 staff 表，为程序提供添加、删除和更新等功能，并尝试将下表中的数据添加到 staff 表中。

6. 建立一个 ContentProvider，用来共享第 4 题所建立的数据库，如下表 11-7 所示。

表 11-7　ContentProvider 共享数据库

_id	name	sex	department	salary
1	Tom	male	computer	5400
2	Einstein	male	computer	4800
3	Lily	female	1.68	5000
4	Warner	male		
5	Napoleon	male		

第 12 章

联系人

☆ 12.1 联系人数据库
☆ 12.2 对联系人的基本操作

本章学习目标：掌握联系人数据库的结构；掌握对联系人的基本操作，包括添加、查询、修改、删除等操作。

12.1 联系人数据库

联系人的数据库文件存放在/data/data/com.android.providers.contacts/databases.contacts2.db 数据库中重要的几张表中。

1. contacts 表

该表保存了所有的手机侧联系人，每个联系人占一行，同时保存了联系人的 ContactID、联系次数、最后一次联系的时间、是否含有号码、是否被添加到收藏夹等信息。

2. raw_contacts 表

该表保存了所有创建过的手机侧联系人，每个联系人占一行，表里有一列标识该联系人是否被删除。该表保存了两个 ID：RawContactID 和 ContactID，从而将 contacts 表和 raw_contacts 表联系起来。该表保存了联系人的 RawContactID、ContactID、联系次数、最后一次联系的时间、是否被添加到收藏夹、显示的名字、用于排序的汉语拼音等信息。

3. mimetypes

该表定义了所有的 MimeTypeID，即联系人的各个字段的唯一标志，如下表 12-1 所示。

表 12-1 mimetypes 表

RecNo	_id	mimetype
1	1	vnd.android.cursor.item/email_v2
2	2	vnd.android.cursor.item/im
3	3	vnd.android.cursor.item/postal-address_v2
4	4	vnd.android.cursor.item/photo
5	5	vnd.android.cursor.item/phone_v2
6	6	vnd.android.cursor.item/name
7	7	vnd.android.cursor.item/organization
8	8	vnd.android.cursor.item/nickname
9	9	vnd.android.cursor.item/group_membership
10	10	vnd.android.cursor.item/website
11	11	vnd.android.cursor.item/note

4. data 表

该表保存了所有创建过的手机侧联系人的所有信息，每个字段占一行。该表保存了两个 ID：MimeTypeID 和 RawContactID，从而将 data 表和 raw_contacts 表联系起来。

联系人的所有信息保存在列 data1 至 data15 中，各列中保存的内容根据 MimeTypeID 的不同而不同。如保存号码(MimeTypeID=5)的那行数据中，data1 列保存号码，data2 列保存号码类型(手机号码/家庭号码/工作号码等)，如下表 12-2 所示。

表 12－2　data 表

mimetype_id	raw_contact_id	is_primary	is_super_primary	data_version	data1	data2
5	1	1	1	2	1	1
6	1	0	0	0	A B	A
6	2	0	0	0	1 2	1
5	3	0	0	0	123	1
5	4	0	0	0	13636431707	2
5	4	0	0	0	05175732549	1
6	4	0	0	1	王刚	刚

12.2　对联系人的基本操作

这里的基本操作只是针对手机侧的联系人，(U)SIM 侧的联系人的操作后续介绍。如果对联系人执行基本操作，我们必须得到许可。方法就是在 AndroidManifest.xml 文件中配置如下权限：

＜uses－permission android:name="android.permission.READ_CONTACTS" /＞
＜uses－permission android:name="android.permission.WRITE_CONTACTS" /＞

12.2.1　读取联系人

分为以下步骤：
(1) 先读取 contacts 表，获取 ContactsID。
(2) 再在 raw_contacts 表中根据 ContactsID 获取 RawContactsID。
(3) 然后就可以在 data 表中根据 RawContactsID 获取该联系人的各数据了。
// 获取用来操作数据的类的对象，对联系人的基本操作都是使用这个对象
ContentResolver cr = getContentResolver();
// 查询 contacts 表的所有记录
Cursor cur = cr.query(ContactsContract.Contacts.CONTENT_URI, null, null, null, null);
// 如果记录不为空
if (cur.getCount() > 0)
{
　　// 游标初始指向查询结果的第一条记录的上方，执行 moveToNext 函数会判断 // 下一条记录是否存在，如果存在，指向下一条记录。否则，返回 false。
　　while (cur.moveToNext())
　　{
　　　　String rawContactsId = "";
　　　　String id = cur.getString(cur.getColumnIndex(ContactsContract.Contacts._ID));
　　　　str += "ID:" + id + "\n";

```
// 读取 rawContactsId
Cursor rawContactsIdCur = cr.query(RawContacts.CONTENT_URI,
        null,
        RawContacts.CONTACT_ID +" = ?",
        new String[]{id}, null);

// 该查询结果一般只返回一条记录,所以我们直接让游标指向第一条记录
if (rawContactsIdCur.moveToFirst())
{
        // 读取第一条记录的 RawContacts._ID 列的值
        rawContactsId = rawContactsIdCur.getString(rawContactsIdCur.getColumnIndex(
            RawContacts._ID));
}
rawContactsIdCur.close();

// 读取号码
If (Integer.parseInt(cur.getString(cur.getColumnIndex(ContactsContract.
    Contacts.HAS_PHONE_NUMBER))) > 0)
{
        // 根据查询 RAW_CONTACT_ID 查询该联系人的号码
        Cursor PhoneCur = cr.query(ContactsContract.CommonDataKinds.Phone.CONTENT_URI,
            null,
            ContactsContract.CommonDataKinds.Phone.RAW_CONTACT_ID +" = ?",
            new String[]{rawContactsId}, null);

        // 上面的 ContactsContract.CommonDataKinds.Phone.CONTENT_URI
        // 可以用下面的 phoneUri 代替
        // Uri phoneUri=Uri.parse("content://com.android.contacts/data/phones");


        // 一个联系人可能有多个号码,需要遍历
        while (PhoneCur.moveToNext())
        {
            // 号获取码
            String number =
        PhoneCur.getString(PhoneCur.getColumnIndex(
                ContactsContract.CommonDataKinds.Phone.NUMBER));
            // 获取号码类型
```

```
            String numberType =
    PhoneCur.getString(PhoneCur.getColumnIndex(
            ContactsContract.CommonDataKinds.Phone.TYPE));
    }
    PhoneCur.close();
}
```

12.2.2 新建联系人

新建联系人时，根据 contacts、raw_contacts 两张表中 ID 的使用情况，自动生成 ContactID 和 RawContactID。Android 源码新建重复姓名的联系人的 ContactID 是不重复的，所以会重复显示。

用下面的代码新建联系人，如果多次新建的联系人的姓名是一样的，生成的 ContactID 也会重复，RawContactID 不会重复，我们在读取联系人的时候可以获取所有同姓名联系人的号码等信息，在显示联系人的时候，重复姓名的联系人的所有字段信息都会合并起来显示为一个联系人。

```
ContentValues values = new ContentValues();
Uri rawContactUri =   getContentResolver().insert(RawContacts.CONTENT_URI,
values);
long rawContactId = ContentUris.parseId(rawContactUri);

// 向 data 表插入姓名数据
if (name ! = "")
{
    values.clear();
    values.put(Data.RAW_CONTACT_ID, rawContactId);
    values.put(Data.MIMETYPE,
StructuredName.CONTENT_ITEM_TYPE);
    values.put(StructuredName.GIVEN_NAME, name);
    getContentResolver().insert(ContactsContract.Data.CONTENT_URI,
values);
}

// 向 data 表插入电话数据
if (phoneNum ! = "")
{
    values.clear();
    values.put(Data.RAW_CONTACT_ID, rawContactId);
    values.put(Data.MIMETYPE, Phone.CONTENT_ITEM_TYPE);
```

```
        values.put(Phone.NUMBER, phoneNum);
        values.put(Phone.TYPE, Phone.TYPE_MOBILE);
        getContentResolver().insert(ContactsContract.Data.CONTENT_URI,
values);
    }
```

12.2.3 删除联系人

Android 帮助文档：When a raw contact is deleted, all of its Data rows as well as StatusUpdates, AggregationExceptions, PhoneLookup rows are deleted automatically.

所以，要删除联系人，我们只需要将 raw_contacts 表中指定 RawContactID 的行删除，其他表中与之关联的数据都会自动删除。

```
    public void delete(long rawContactId)
    {
        getContentResolver().delete(ContentUris.withAppendedId(RawContacts.CONTENT_URI, rawContactId), null, null);
    }
```

12.2.4 更新联系人

联系人的所有信息都是保存在 data 表中，所以要更新联系人，我们只需要根据 RawContactID 和 MIMETYPE 修改 data 表中的内容。

```
    ContentValues values = new ContentValues();
    values.put(Phone.NUMBER, "123");
    values.put(Phone.TYPE, Phone.TYPE_MOBILE);

    String Where = ContactsContract.Data.RAW_CONTACT_ID + " = ? AND " +
ContactsContract.Data.MIMETYPE + " = ?";
    String[] WhereParams = new String[]{"5", Phone.CONTENT_ITEM_TYPE};

    getContentResolver().update(ContactsContract.Data.CONTENT_URI, values,
Where, WhereParams);
```

习题与思考题

1. 使用 DDMS 等工具导出，或者在已经 Root 过的 Android 手机上使用 RE 管理器等软件，打开并观察 Android 联系人数据库，体会 Android 联系人数据库的设计思想。

2. 编写一个简单的联系人管理应用，可以实现 Android 手机联系人的添加、查找、修改、删除等操作。

第 13 章　Android 图形开发

☆ 13.1　Drawable 对象
☆ 13.2　Bitmap 对象
☆ 13.3　Animation 对象

本章学习目标:Drawable 对象的使用;Bitmap 对象的使用;Animation 对象的使用;Canvas 的使用。

13.1 Drawable 对象

数据包:android.content.res
主要类:Resources
主要接口:
int getColor(int id)　　//获取 RES 中的资源
Drawable getDrawable(int id)
String getString(int id)
InputStream openRawResource(int id)　　//获取资源数据流
获取当前 resource 的方法:
Resources r = this.getContext().getResources();
包:android.graphics.drawable
主要类:Drawable

Drawable 就是一个可画的对象,其可能是一张位图(BitmapDrawable),也可能是一个图形(ShapeDrawable),还有可能是一个图层(LayerDrawable)。我们根据画图的需求,创建相应的可画对象,就可以将这个可画对象当作一块"画布(Canvas)",在其上面操作可画对象,并最终将这种可画对象显示在画布上。

通过 Drawable 子类 ShapeDrawable 的简单例子了解一下 Drawable,运行结果如图 13-1 所示。

图 13-1　ShapeDrawable 的运行结果

代码如下:
```
public class testView extends View {
  private ShapeDrawable mDrawable;
  public testView(Context context) {
    super(context);
    int x = 10;
    int y = 10;
    int width = 300;
```

第 13 章 Android 图形开发

```
        int height = 50;
        mDrawable = new ShapeDrawable(new OvalShape());
        mDrawable.getPaint().setColor(0xff74AC23);
        Drawable.setBounds(x, y, x + width, y + height);
    }
    protected void onDraw(Canvas canvas)
        super.onDraw(canvas);
        canvas.drawColor(Color.WHITE);//画白色背景
        mDrawable.draw(canvas)
```

代码说明：创建一个 OvalShape（椭圆），使用刚创建的 OvalShape 构造一个 ShapeDrawable 对象 mDrawable；设置 mDrawable 的颜色，设置 mDrawable 的大小；将 mDrawable 画在 testView 的画布上。

主要作用：

在 XML 中定义各种动画，然后把 XML 当作 Drawable 资源来读取，通过 Drawable 显示动画。下面举个使用 TransitionDrawable 的例子，创建一个 Android 工程，然后在这个工程的基础上修改。

（1）去掉 layout/main.xml 中的 TextView，增加 ImagView，内容如下：

```
<ImageView
android:layout_width="wrap_content"
android:layout_height="wrap_content"
android:tint="#55ff0000"
android:src="@drawable/my_image"/>
```

（2）创建一个 XML 文件，命名为 expand_collapse.xml，内容如下：

```
<?xml version="1.0" encoding="UTF-8"?>
<transition xmlns:android="http://schemas.android.com/apk/res/android">
<item android:drawable="@drawable/image_expand"/>
<item android:drawable="@drawable/image_collapse"/>
</transition>
```

需要 3 张 png 图片，存放到 res\drawable 目录下，3 张图片分别命名为：my_image.png、image_expand.png、image_collapse.png。

（3）修改 Activity 中的代码，内容如下：

```
LinearLayout mLinearLayout;
protected void onCreate(Bundle savedInstanceState) {
    super.onCreate(savedInstanceState);
    mLinearLayout = new LinearLayout(this);
    ImageView i = new ImageView(this);
    i.setAdjustViewBounds(true);
    i.setLayoutParams(new Gallery.LayoutParams(LayoutParams.WRAP_CONTENT,
    LayoutParams.WRAP_CONTENT));
    mLinearLayout.addView(i);
```

```
        setContentView(mLinearLayout);
        Resources res = getResources();
        TransitionDrawable transition =
        (TransitionDrawable) res.getDrawable(R.drawable.expand_collapse);
        i.setImageDrawable(transition);
        transition.startTransition(10000);
    }
```

运行结果如图 13-2,13-3,13-4 所示。

图 13-2 显示图片 image_expand.png

图 13-3 显示图片 my_image.png

图 13-4 显示图片 image_collapse.png

屏幕上动画显示的是：从图片 image_expand.png 过渡到 image_collapse.png,也就是我们在 expand_collapse.xml 中定义的一个 transition 动画。

主要接口：

BitmapDrawable()

BitmapDrawable(Bitmap bitmap)

BitmapDrawable(String filepath)

BitmapDrawable(InputStream is)

final Bitmap getBitmap()

final Paint getPaint()

void draw(Canvas canvas)

在 BitmapDrawable 中我们就看到位图的具体操作,在仔细看下 BitmapDrawable 的构造函数,我们就会发现与 Resource 中的 openRawResource()接口是相对应的,就可以通过以下方法来获取位图：

Resources r = this.getContext().getResources();
Inputstream is = r.openRawResource(R.drawable.my_background_image);
BitmapDrawable bmpDraw = new BitmapDrawable(is);
Bitmap bmp = bmpDraw.getBitmap();

包：android.graphics

Android SDK 中的简介：The Paint class holds the style and color information about how to draw geometries, text and bitmaps. 主要就是定义：画刷的样式,画笔的大小/颜色等。

Typeface

包：android.graphics

Android SDK 中的简介：The Typeface class specifies the typeface and intrinsic style of a font. 主要就是定义：字体。

13.2 Bitmap 对象

位图是我们开发中最常用的资源,毕竟一个漂亮的界面对用户是最有吸引力的。按照对位图的操作,分为以下几个功能分别介绍：

从资源中获取位图,通过 Resource 的函数 InputStream openRawResource(int id)获取得到资源文件的数据流后,获取 Bitmap。

1. 使用 BitmapDrawable

使用 BitmapDrawable (InputStream is)构造一个 BitmapDrawabl,使用 BitmapDrawable 类的 getBitmap()获取得到位图。

2. 使用 BitmapFactory

使用 BitmapFactory 类 decodeStream(InputStream is)解码位图资源,获取位图。

(1) 在 Android SDK 中说明可以支持的图片格式如下：png（preferred），jpg（acceptable），gif（discouraged）,虽然 bmp 格式没有明确说明,但是在 Android SDK Support Media Format 中是明确说明了。

(2) 获取位图信息

要获取位图信息,比如位图大小、是否包含透明度、颜色格式等,获取得到 Bitmap 就迎刃而解了。这些信息在 Bitmap 的函数中可以轻松获取到,例如：

getHeight();
getWidth();

压缩方法：compress（Bitmap.CompressFormat format, int quality, OutputStream stream)

(3) 显示位图

显示位图需要使用核心类 Canvas,可以直接通过 Canvas 类的 drawBirmap()显示位图,或者借助于 BitmapDrawable 来将 Bitmap 绘制到 Canvas。下面举例如何获取 Canvas。

例子中包含两个类,用于画面显示的 testActivity 及自定义 View 类 testView。代码如下：

```java
public class testActivity extends Activity {
private testView mTestview;
    /** Called when the activity is first created. */
    @Override
    public void onCreate(Bundle savedInstanceState) {
        super.onCreate(savedInstanceState);
        setContentView(R.layout.main);
        mTestview = (testView) findViewById(R.id.testView);
        mTestview.initBitmap(320,240,0xcccccc);
    }
}
public class testView extends View {
private Bitmap  mbmpTest=null;
private final Paint mPaint = new Paint();
private final String mstrTitle="感受 Android 带给我们的新体验";
public testView(Context context, AttributeSet attrs, int defStyle)
{
    super(context, attrs, defStyle);
    mPaint.setColor(Color.GREEN);
}
public testView(Context context, AttributeSet attrs)
{
    super(context, attrs);
    mPaint.setColor(Color.GREEN);
}
public boolean initBitmap(int w,int h,int c)
{
mbmpTest = Bitmap.createBitmap(w,h, Config.ARGB_8888);
Canvas canvas = new Canvas(mbmpTest);
canvas.drawColor(Color.WHITE);
Paint p = new Paint();
String familyName = "宋体";
Typeface font = Typeface.create(familyName,Typeface.BOLD);
p.setColor(Color.RED);
p.setTypeface(font);
p.setTextSize(22);
canvas.drawText(mstrTitle,0,100,p);
```

```
        return true;
}
@Override
    public void onDraw(Canvas canvas)
{
super.onDraw(canvas);

if(mbmpTest! =null)
{
            Matrix matrix = new Matrix();
            matrix.setRotate(90,120,120);
            canvas.drawBitmap(mbmpTest, matrix, mPaint);
}
}
```

main.xml

```
<?xml version="1.0" encoding="utf-8" ?>
<FrameLayout    xmlns:android="http://schemas.android.com/apk/res/android" android:layout_width="fill_parent" android:layout_height="fill_parent">
        <testView.moandroid.testView    android:id="@+id/testView" android:layout_width="fill_parent" android:layout_height="fill_parent" tileSize="12" />
</FrameLayout>
```

testView 例子介绍:其包含 2 个类 testActivity,testView;testActivity 继承与 Activity,testView 继承与 View。这个例子就是将 testView 直接作为 testActivity 的窗口,这样我们就可以直接在 testView 画图了。具体实现的方法请大家参考 testActivity 的 onCreate()中的代码,以及 layout\main.xml 中的设置。在 testView 的 onDraw()直接画图,结果在例子程序运行后就可以直接在界面上显示了。

(4) 缩放

① 将一个位图按照需求重画一遍,画后的位图就是我们需要的了,与位图的显示几乎一样:drawBitmap(Bitmap bitmap, Rect src, Rect dst, Paint paint)。

② 在原有位图的基础上,缩放原位图,创建一个新的位图:CreateBitmap(Bitmap source, int x, int y, int width, int height, Matrix m, boolean filter)。

③ 借助 Canvas 的 scale(float sx, float sy)(Preconcat the current matrix with the specified scale.),不过要注意此时整个画布都缩放了。

④ 借助 Matrix:

Bitmap bmp = BitmapFactory.decodeResource(getResources(), R.drawable.pic180);

Matrix matrix=new Matrix();

matrix.postScale(0.2f, 0.2f);

Bitmap dstbmp=Bitmap.createBitmap(bmp,0,0,bmp.getWidth(),

bmp.getHeight(),matrix,true);
canvas.drawColor(Color.BLACK);
canvas.drawBitmap(dstbmp,10,10,null);

(5) 旋转

位图的旋转也可以借助 Matrix 或者 Canvas 来实现。
Bitmap bmp = BitmapFactory.decodeResource(getResources(), R.drawable.pic180);
Matrix matrix=new Matrix();
matrix.postScale(0.8f, 0.8f);
matrix.postRotate(45);
Bitmap dstbmp=Bitmap.createBitmap(bmp,0,0,bmp.getWidth(),
bmp.getHeight(),matrix,true);
canvas.drawColor(Color.BLACK);
canvas.drawBitmap(dstbmp,10,10,null);

onDraw 方法会传入一个 Canvas 对象，它是你用来绘制控件视觉界面的画布。在 onDraw 方法里，我们经常会用到 save 和 restore 方法，这两个方法的作用如下：

save：用来保存 Canvas 的状态。save 之后，可以调用 Canvas 的平移、放缩、旋转、错切、裁剪等操作。

restore：用来恢复 Canvas 之前保存的状态。防止 save 后对 Canvas 执行的操作对后续的绘制有影响。

save 和 restore 要配对使用（restore 可以比 save 少，但不能多），如果 restore 调用次数比 save 多，会引发 Error。save 和 restore 之间，往往夹杂的是对 Canvas 的特殊操作。

例如，我们先想在画布上绘制一个右向的三角箭头，当然，可以直接绘制，也可以先把画布旋转 90°，画一个向上的箭头，再旋转回来（这种旋转操作对于画圆周上的标记非常有用）；然后，我们想在右下角有个 20 像素的圆。

效果如图 13-5 和图 13-6 所示。

图 13-5 绘制一个向右的三角箭头　　　　图 13-6 旋转 90°后的图片

13.3 Animation 对象

Android 中主要使用了两种 Animation：

(1) Tween Animation，通过对场景里的对象不断做图像变换(平移、缩放、旋转)产生动画效果。

(2) Frame Animation，顺序播放事先做好的图像，跟电影类似。

1. Tween Animation

(1) Tween Animation 有四种类型。

① Alpha：渐变透明度动画效果；

② Scale：渐变尺寸伸缩动画效果；

③ Translate：画面转换位置移动动画效果；

④ Rotate：画面转换位置移动动画效果。

Duration[long]属性为动画持续时间，时间以毫秒为单位。fillAfter[boolean]当设置为 true，该动画转化在动画结束后被应用。fillBefore[boolean]当设置为 true，该动画转化在动画开始前被应用。

(2) interpolator：指定一个动画的插入器，

有一些常见的插入器如下。

accelerate_decelerate_interpolator：加速-减速动画插入器。

accelerate_interpolator：加速-动画插入器。

decelerate_interpolator：减速-动画插入器。

repeatCount[int]，动画的重复次数。

RepeatMode[int]，定义重复的行为。

startOffset[long]，动画之间的时间间隔，从上次动画停多少时间开始执行下个动画；

Adjustment[int]，定义动画的 Z Order 的改变。0：保持 Z Order 不变；1：保持在最上层；−1：保持在最下层。

Alpha 节点如表 13-1 所示。

表 13-1 Alpha 节点

XML 节点	功能说明
alpha	渐变透明度动画效果

```
<alpha
android:fromAlpha="0.1"
android:toAlpha="1.0"
android:duration="3000"/>
```

fromAlpha	属性为动画起始时透明度	0.0 表示完全透明
		1.0 表示完全不透明
toAlpha	属性为动画结束时透明度	以上值取 0.0—1.0 之间的 float 数据类型的数字

Scale 节点如表 13-2 所示。

表 13-2　Scale 节点

scale	渐变尺寸伸缩动画效果
<scale android:interpolator="@android:anim/accelerate_decelerate_interpolator" android:fromXScale="0.0" android:toXScale="1.4" android:fromYScale="0.0" android:toYScale="1.4" android:pivotX="50%" android:pivotY="50%" android:fillAfter="false" android:startOffset="700" android:duration="700" android:repeatCount="10/>"	
fromXScale[float] fromYScale[float]	为动画起始时,X、Y 坐标上的伸缩尺寸	0.0 表示收缩到没有 1.0 表示正常无伸缩
toXScale[float] toYScale[float]	为动画结束时,X、Y 坐标上的伸缩尺寸	值小于 1.0 表示收缩 值大于 1.0 表示放大
pivotX[float] pivotY[float]	为动画相对于物件的 X、Y 坐标的开始位置	属性值说明:从 0%—100% 中取值,50% 为物件的 X 或 Y 方向坐标上的中点位置

Translate 节点如表 13-3 所示。

表 13-3　Translate 节点

translate	画面转换位置移动动画效果
<translate android:fromXDelta="30" android:toXDelta="-80" android:fromYDelta="30" android:toYDelta="300" android:duration="2000"/>	
fromXDelta toXDelta	为动画、结束起始时 X 坐标上的位置	
fromYDelta toYDelta	为动画、结束起始时 Y 坐标上的位置	

Rotate 节点如表 13-4 所示。

表 13-4　Rotate 节点

rotate	画面转移旋转动画效果
<rotate android:interpolator="@android:anim/accelerate_decelerate_interpolator" android:fromDegrees="0" android:toDegrees="+350" android:pivotX="50%" android:pivotY="50%" android:duration="3000"/>	

续 表

rotate	画面转移旋转动画效果
fromDegrees	为动画起始时物件的角度
toDegrees	属性为动画结束时物件旋转的角度可以大于360度
pivotX pivotY	为动画相对于物件的X、Y坐标的开始位

说明：
当角度为负数——表示逆时针旋转
当角度为正数——表示顺时针旋转
（负数 from——to 正数：顺时针旋转）
（负数 from——to 负数：逆时针旋转）
（正数 from——to 正数：顺时针旋转）
（正数 from——to 负数：逆时针旋转）

说明：以上两个属性值从 0%—100% 中取值
50%为物件的X或Y方向坐标上的中点位置

Scale 动画的例子：

（1）创建 Android 工程；

（2）导入一张图片资源，将 res\layout\main.xml 中的 TextView 取代为 ImageView；在 res 下创建新的文件夹 anim，并在此文件夹下面定义 Animation XML 文件；修改 OnCreate() 中的代码，显示动画资源。

关键代码解析：

// main.xml 中的 ImageView

ImageView spaceshipImage = (ImageView) findViewById(R.id.spaceshipImage);

// 加载动画

Animation hyperspaceJumpAnimation =
AnimationUtils.loadAnimation(this, R.anim.hyperspace_jump);

// 使用 ImageView 显示动画

spaceshipImage.startAnimation(hyperspaceJumpAnimation);

Tween Animation 通过对 View 的内容完成一系列的图形变换（包括平移、缩放、旋转、改变透明度）来实现动画效果。具体来讲，预先定义一组指令，这些指令指定了图形变换的类型、触发时间、持续时间。这些指令可以是以 XML 文件方式定义，也可以是以源代码方式定义。程序沿着时间线执行这些指令就可以实现动画效果。

动画控制：

对 interpolator 的说明，interpolator 定义一个动画的变化率(the rate of change)。这使得基本的动画效果(alpha，scale，translate，rotate)得以加速、减速、重复等。

动画的进度使用 Interpolator 控制。Interpolator 定义了动画的变化速度，可以实现匀速、正加速、负加速、无规则变加速等。Interpolator 是基类，封装了所有 Interpolator 的共同方法，它只有一个方法，即 getInterpolation (float input)。

Interpolator 的几个子类如表 13-5 所示。

表 13-5 Interpolator 的几个子类

AccelerateDecelerateInterpolator	在动画开始与介绍的地方速率改变比较慢，在中间的时候加速
AccelerateInterpolator	在动画开始的地方速率改变比较慢，然后开始加速
CycleInterpolator	动画循环播放特定的次数，速率改变沿着正弦曲线

续 表

DecelerateInterpolator	在动画开始的地方速率改变比较慢,然后开始减速
LinearInterpolator	在动画的以均匀的速率改变

动画的运行的两种模式:

(1) 独占模式,即程序主线程进入一个循环,根据动画指令不断刷新屏幕,直到动画结束。

(2) 中断模式,即有单独一个线程对时间计数,每隔一定的时间向主线程发通知,主线程接到通知后更新屏幕。

2. Frame Animation

Frame Animation 的定义:Frame Animation 可以在 XML Resource 定义(还是存放到 res\anim 文件夹下),也可以使用 AnimationDrawable 中的 API 定义。

animation-list 根节点中包含多个 item 子节点,每个 item 节点定义一帧动画:当前帧的 drawable 资源和当前帧持续的时间。

XML 属性和说明如表 13-6 所示。

表 13-6 XML 属性和说明

XML 属性	说 明
drawable	当前帧引用的 drawable 资源
duration	当前帧显示的时间(毫秒为单位)
oneshot	如果为 true,表示动画只播放一次停止在最后一帧上,如果设置为 false 表示动画循环播放。
variablePadding	If true, allows the drawable's padding to change based on the current state that is selected.
visible	规定 drawable 的初始可见性,默认为 flase;

下面就给个具体的 XML 例子,来定义一帧一帧的动画:

```
<animation-list xmlns:android="http://schemas.android.com/apk/res/android" android:oneshot="true">
<item android:drawable="@drawable/rocket_thrust1" android:duration="200" />
<item android:drawable="@drawable/rocket_thrust2" android:duration="200" />
<item android:drawable="@drawable/rocket_thrust3" android:duration="200" />
</animation-list>
```

在 OnCreate() 中增加如下代码:

```
ImageView rocketImage = (ImageView) findViewById(R.id.rocket_image);
rocketImage.setBackgroundResource(R.anim.rocket_thrust);
   rocketAnimation = (AnimationDrawable) rocketImage.getBackground();
```

最后还需要增加启动动画的代码:

```
public boolean onTouchEvent(MotionEvent event) {
if (event.getAction() == MotionEvent.ACTION_DOWN) {
rocketAnimation.start();
return true;
```

}
return super.onTouchEvent(event);
}

注意:代码运行的结果想必大家已经知道(3张图片按照顺序的播放一次),不过有一点需要强调的是:启动 Frame Animation 动画的代码 rocketAnimation.start();不能在 OnCreate()中,因为在 OnCreate()中 AnimationDrawable 还没有完全的与 ImageView 绑定,在 OnCreate()中启动动画,就只能看到第一张图片。

AnimationDrawable 如表 13-7 所示。

表 13-7 **AnimationDrawable**

获取、设置动画的属性	
int getDuration()	获取动画的时长
int getNumberOfFrames()	获取动画的帧数
boolean isOneShot() Void setOneShot(boolean oneshot)	获取 oneshot 属性 设置 oneshot 属性
void inflate(Resurce r, XmlPullParser P, AttributeSet attrs)	
增加、获取帧动画	
Drawable getFrame(int index)	获取某帧的 Drawable 资源
void addFrame(Drawable frame, intduration)	为当前动画增加帧(资源,持续时长)
动画控制	
void start()	开始动画
void run()	外界不能直接调用,使用 start()替代
boolean isRunning()	当前动画是否在运行
void stop()	停止当前动画

习题与思考题

1. 在 Android 中,Drawable 对象和 Bitmap 对象有哪些相似的地方?它们又有哪些不同?在使用时分别要注意什么?

2. 简述 Animation 实现动画的两种方式,它们在思路上有何区别,各有哪些优缺点?

3. 编写一个简单的 Android 应用,在界面上使用 Canvas 绘制 Drawble 和 Bitmap 对象,并实现 Animation 动画。

第 14 章 SurfaceView

☆ 14.1 SurfaceView 简介
☆ 14.2 自定义 SurfaceView
☆ 14.3 SurfaceView 的多线程

本章学习目标：掌握 SurfaceView 的基本用法；掌握自定义 SurfaceView 的方法；使用多线程控制 SurfaceView。

14.1 SurfaceView 简介

SurfaceView 由于可以直接从内存或者 DMA 等硬件接口取得图像数据，因此是个非常重要的绘图容器。SurfaceView 的用法有很多，写法也层出不穷，例如继承 SurfaceView 类，或者继承 SurfaceHolder.Callback 类等，这个可以根据功能实际需要自己选择。第一个例子我们直接在普通的用户界面调用 SurfaceHolder 的 lockCanvas 和 unlockCanvasAndPost。

将正弦曲线绘画在 SurfaceView 上，如图 14-1 所示。

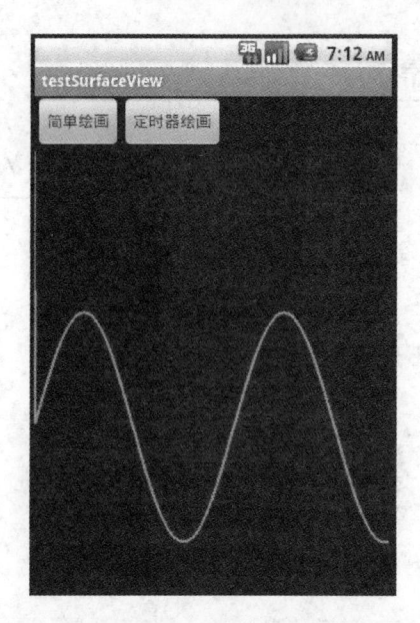

图 14-1 将正弦曲线绘画在 SurfaceView 上

Main.xml 文件的代码如下。

```
<?xml version = "1.0" encoding = "utf-8"?>
<LinearLayout xmlns:android = "http://schemas.android.com/apk/res/android"
   android:layout_width = "fill_parent" android:layout_height = "fill_parent"
   android:orientation = "vertical">
<LinearLayout android:id = "@+id/LinearLayout01"
   android:layout_width = "wrap_content" android:layout_height = "wrap_content">
   <Button android:id = "@+id/Button01" android:layout_width = "wrap_content"
      android:layout_height = "wrap_content" android:text = "简单绘画">
   </Button>
   <Button android:id = "@+id/Button02" android:layout_width = "wrap_content"
      android:layout_height = "wrap_content" android:text = "定时器绘画">
   </Button>
```

</ LinearLayout >
< SurfaceView android:id = "@+id/SurfaceView01"
    android:layout_width = "fill_parent" android:layout_height = "fill_parent" ></ SurfaceView >
</ LinearLayout >

需要导入的包：
```java
import java.util.Timer;
import java.util.TimerTask;
import android.app.Activity;
import android.graphics.Canvas;
import android.graphics.Color;
import android.graphics.Paint;
import android.graphics.Rect;
import android.os.Bundle;
import android.util.Log;
import android.view.SurfaceHolder;
import android.view.SurfaceView;
import android.view.View;
import android.widget.Button;
```

Activity：
```java
public class testSurfaceView  extends  Activity {
    /** Called when the activity is first created. */
    Button btnSimpleDraw, btnTimerDraw;
    SurfaceView sfv;
    SurfaceHolder sfh;

    private Timer mTimer;
    private MyTimerTask  mTimerTask;
    int Y_axis[]; //保存正弦波的Y轴上的点
    int centerY; //中心线
    int oldX,oldY; //上一个XY点
    int currentX; //当前绘制到的X轴上的点
    @Override
    public void onCreate(Bundle savedInstanceState) {
        super.onCreate(savedInstanceState);
        setContentView(R.layout.main);

        btnSimpleDraw = (Button) this.findViewById(R.id.Button01);
```

```java
btnTimerDraw = (Button)this.findViewById(R.id.Button02);
btnSimpleDraw.setOnClickListener(new ClickEvent());
btnTimerDraw.setOnClickListener(new ClickEvent());
sfv = (SurfaceView)this.findViewById(R.id.SurfaceView01);
sfh = sfv.getHolder();

//动态绘制正弦波的定时器
mTimer = new Timer();
mTimerTask = new MyTimerTask();

// 初始化y轴数据
centerY = (getWindowManager().getDefaultDisplay().getHeight() - sfv
        .getTop()) / 2;
Y_axis = new    int[getWindowManager().getDefaultDisplay().getWidth()];
for (int i=1; i< Y_axis.length; i++) { //计算正弦波
    Y_axis[i-1] = centerY
        - (int)(100 * Math.sin(i * 2 * Math.PI / 180));
}
}

class ClickEvent implements   View.OnClickListener {
    @Override
    public void onClick(View v) {

        if(v == btnSimpleDraw) {
            SimpleDraw(Y_axis.length-1); //直接绘制正弦波

        } else  if (v == btnTimerDraw) {
            oldY = centerY;
            mTimer.schedule(mTimerTask, 0, 5); //动态绘制正弦波
        }
    }
}

class MyTimerTask extends TimerTask {
    @Override
    public void run() {
        SimpleDraw(currentX);
        currentX++; //往前进
```

```
            if (currentX == Y_axis.length - 1){ //如果到了终点,则清屏重来
                ClearDraw();
                currentX = 0;
                oldY = centerY;
            }
        }
    }

    void SimpleDraw( int  length){
        if (length ==0 )
            oldX = 0 ;
        Canvas canvas = sfh.lockCanvas(new  Rect(oldX,0, oldX + length,
            getWindowManager().getDefaultDisplay().getHeight()));// 关键:获取画布
        Log.i("Canvas:" ,
            String.valueOf(oldX) + ","+ String.valueOf(oldX + length));

        Paint mPaint = new  Paint();
        mPaint.setColor(Color.GREEN);// 画笔为绿色
        mPaint.setStrokeWidth(2 ); // 设置画笔粗细

        int y;
        for ( int  i = oldX + 1 ; i < length; i++){ // 绘画正弦波
            y = Y_axis[i - 1 ];
            canvas.drawLine(oldX, oldY, i, y, mPaint);
            oldX = i;
            oldY = y;
        }
        sfh.unlockCanvasAndPost(canvas);// 解锁画布,提交画好的图像
    }

    void ClearDraw(){
        Canvas canvas = sfh.lockCanvas(null );
        canvas.drawColor(Color.BLACK);// 清除画布
        sfh.unlockCanvasAndPost(canvas);
    }
}
```

14.2 自定义 SurfaceView

继承 SurfaceView 类实现 SurfaceHolder.Callback。SurfaceHolder.Callback 在底层的 Surface 发生变化的时候通知 view。

1. SurfaceHolder.Callback 接口

surfaceCreated(SurfaceHolder holder)：当 Surface 第一次创建后会立即调用该函数。程序可以在该函数中做些和绘制界面相关的初始化工作，一般情况下都是在另外的线程来绘制界面，所以不要在这个函数中绘制 Surface。

surfaceChanged(SurfaceHolder holder，int format，int width，int height)：当 Surface 的状态（大小和格式）发生变化的时候会调用该函数，在 surfaceCreated 调用后该函数至少会被调用一次。

surfaceDestroyed(SurfaceHolder holder)：当 Surface 被摧毁前会调用该函数，该函数被调用后就不能继续使用 Surface 了，一般在该函数中来清理使用的资源。

2. SurfaceHolder

（1）取得方法

通过 SurfaceView 的 getHolder() 函数可以获取 SurfaceHolder 对象，Surface 就在 SurfaceHolder 对象内。

（2）使用方法

Surface 保存了当前窗口的像素数据，但是在使用过程中是不直接和 Surface 打交道的，由 SurfaceHolder 的 Canvas lockCanvas() 或则 Canvas lockCanvas(Rect dirty) 函数来获取 Canvas 对象，通过在 Canvas 上绘制内容来修改 Surface 中的数据。

（3）SurfaceHolder 注意事项：

① Surface 不可编辑或则尚未创建调用 getHolder() 函数会返回 null。

② 在 unlockCanvas() 和 lockCanvas() 中 Surface 的内容是不缓存的，所以需要完全重绘 Surface 的内容，为了提高效率只重绘变化的部分则可以调用 lockCanvas(Rect dirty) 函数来指定一个 dirty 区域，这样该区域外的内容会缓存起来。

（4）类型设置

在 Canvas 中绘制完成后，调用函数 unlockCanvasAndPost(Canvas canvas) 来通知系统 Surface 已经绘制完成，这样系统会把绘制完的内容显示出来。为了充分利用不同平台的资源，发挥平台的最优效果可以通过 SurfaceHolder 的 setType 函数来设置绘制的类型，目前接收如下的参数。

SURFACE_TYPE_NORMAL：用 RAM 缓存原生数据的普通 Surface。

SURFACE_TYPE_HARDWARE：适用于 DMA(Direct memory access)引擎和硬件加速的 Surface。

SURFACE_TYPE_GPU：适用于 GPU 加速的 Surface。

SURFACE_TYPE_PUSH_BUFFERS：表明该 Surface 不包含原生数据，Surface 用到的数据由其他对象提供。例如 Camera 中得图像预览。

3. 使用 Canvas 绘制 SurfaceView 的框架

说明：在 Surface View 控件中创建了一个新的由 Thread 派生的类，所有的 UI 更新都是在这个新类中处理。

代码如下：

```
import android.content.Context;
import android.graphics.Canvas;
import android.view.SurfaceHolder;
import android.view.SurfaceView;
public class MySurfaceView extends SurfaceView implements SurfaceHolder.Callback {
  private SurfaceHolder holder;
  private MySurfaceViewThread mySurfaceViewThread;
  private boolean hasSurface;
  MySurfaceView(Context context) {
      super(context);
      init();
  }
  public void resume() {
    //创建和启动图像更新线程
    if (mySurfaceViewThread == null) {
      mySurfaceViewThread = new MySurfaceViewThread();
      if (hasSurface == true)
        mySurfaceViewThread.start();
    }
  }
  public void pause() {
    // 杀死图像更新线程
    if (mySurfaceViewThread != null) {
      mySurfaceViewThread.requestExitAndWait();
      mySurfaceViewThread = null;
    }
  }
  public void surfaceCreated(SurfaceHolder holder) {
    hasSurface = true;
    if (mySurfaceViewThread != null)
      mySurfaceViewThread.start();
  }
  public void surfaceDestroyed(SurfaceHolder holder) {
    hasSurface = false;
    pause();
```

```
  }
  public void surfaceChanged(SurfaceHolder holder, int format, int w, int h) {
    if (mySurfaceViewThread ! = null)
      mySurfaceViewThread.onWindowResize(w, h);
  }
  class MySurfaceViewThread extends Thread {
    private boolean done;
    MySurfaceViewThread() {
      super();
      done=false;
    }
    @Override
    public void run() {
      SurfaceHolder surfaceHolder = holder;
      // 重复绘图循环,直到线程停止
      while (! done) {    // 锁定 surface,并返回到要绘图的 Canvas
        Canvas canvas=surfaceHolder.lockCanvas();
        // 待实现:在 Canvas 上绘图

        // 解锁 Canvas,并渲染当前图像
        surfaceHolder.unlockCanvasAndPost(canvas);
      }
    }
    public void requestExitAndWait() {
      // 把这个线程标记为完成,并合并到主程序线程
      done=true;
      try {
        join();
      } catch (InterruptedException ex) { }
    }
    public void onWindowResize(int w, int h) {
      // 处理可用的屏幕尺寸的改变
    }
  }
}
```

14.3　SurfaceView 的多线程

SurfaceView 使用多线程在游戏中的作用:

(1) 资源读取与使用分离，便于异步加载。
(2) 提高效率。
(3) 防止同步处理时的阻塞。
(4) 游戏中可以防止动画卡顿闪烁。

Mainl.xml 文件的代码如下。

```xml
<?xml version="1.0" encoding="utf-8"?>
<LinearLayout xmlns:android="http://schemas.android.com/apk/res/android"
    android:layout_width="fill_parent" android:layout_height="fill_parent"
    android:orientation="vertical">
    <LinearLayout android:id="@+id/LinearLayout01"
        android:layout_width="wrap_content" android:layout_height="wrap_content">
        <Button android:id="@+id/Button01" android:layout_width="wrap_content"
            android:layout_height="wrap_content" android:text="单个独立线程">
        </Button>
        <Button android:id="@+id/Button02" android:layout_width="wrap_content"
            android:layout_height="wrap_content" android:text="两个独立线程">
        </Button>
    </LinearLayout>
    <SurfaceView android:id="@+id/SurfaceView01"
        android:layout_width="fill_parent" android:layout_height="fill_parent">
    </SurfaceView>
</LinearLayout>
```

引入包：

```java
import java.lang.reflect.Field;
import java.util.ArrayList;
import android.app.Activity;
import android.graphics.Bitmap;
import android.graphics.BitmapFactory;
import android.graphics.Canvas;
import android.graphics.Paint;
import android.graphics.Rect;
import android.os.Bundle;
import android.util.Log;
import android.view.SurfaceHolder;
import android.view.SurfaceView;
import android.view.View;
```

```java
import   android.widget.Button;
public class testSurfaceView extends Activity {
    /** Called when the activity is first created. */
    Button btnSingleThread, btnDoubleThread;
    SurfaceView sfv;
    SurfaceHolder sfh;
    ArrayList<Integer> imgList = new ArrayList<Integer>();
    int imgWidth, imgHeight;
    Bitmap bitmap;//独立线程读取,独立线程绘图
    @Override
    public void onCreate(Bundle savedInstanceState) {
        super.onCreate(savedInstanceState);
        setContentView(R.layout.main);
        btnSingleThread = (Button) this.findViewById(R.id.Button01);
        btnDoubleThread = (Button) this.findViewById(R.id.Button02);
        btnSingleThread.setOnClickListener(new ClickEvent());
        btnDoubleThread.setOnClickListener(new ClickEvent());
        sfv = (SurfaceView) this.findViewById(R.id.SurfaceView01);
        sfh = sfv.getHolder();
        sfh.addCallback(new MyCallBack());// 自动运行 surfaceCreated 以及
        surfaceChanged
    }
    class ClickEvent implements View.OnClickListener {
        @Override
        public void onClick(View v) {
            if(v == btnSingleThread) {
                new Load_DrawImage(0, 0).start();//开一条线程读取并绘图
            } else if(v == btnDoubleThread) {
                new LoadImage().start();//开一条线程读取
                new DrawImage(imgWidth +10, 0).start();//开一条线程绘图
            }
        }
    }
    class MyCallBack implements SurfaceHolder.Callback {
        @Override
        public void surfaceChanged(SurfaceHolder holder, int format, int width,
            int height) {
            Log.i("Surface:", "Change");
```

```java
        }
    @Override
    public void surfaceCreated(SurfaceHolder holder) {
        Log.i("Surface:","Create");// 用反射机制来获取资源中的图片 ID 和尺寸
        Field[] fields = R.drawable.class.getDeclaredFields();
        for (Field field : fields) {
          if (!"icon".equals(field.getName())) // 除了 icon 之外的图片
          {
            int index =0;
            try{
              index = field.getInt(R.drawable.class);
            } catch (IllegalArgumentException e) {
              // TODO Auto-generated catch block
              e.printStackTrace();
            } catch (IllegalAccessException e) {
              // TODO Auto-generated catch block
              e.printStackTrace();
            }
            // 保存图片 ID
            imgList.add(index);
          }
        }
        // 取得图像大小
        Bitmap bmImg = BitmapFactory.decodeResource(getResources(),
            imgList.get(0));
        imgWidth = bmImg.getWidth();
        imgHeight = bmImg.getHeight();
    }
    @Override
    public void surfaceDestroyed(SurfaceHolder holder) {
        Log.i("Surface:", "Destroy");
    }
  }
  /*
   * 读取并显示图片的线程
   */
  class Load_DrawImage extends Thread {
    int x, y;
```

```
        int imgIndex =0 ;
        public Load_DrawImage( int x,   int y) {
           this . x = x;
           this . y = y;
        }
        public void run() {
           while(true ) {
              Canvas c = sfh. lockCanvas(new Rect(this . x,    this . y,   this . x
                  + imgWidth, this . y + imgHeight));
              Bitmap bmImg = BitmapFactory. decodeResource(getResources(),
                  imgList. get(imgIndex));
              c. drawBitmap(bmImg, this . x,   this . y, new Paint());
              imgIndex++;
              if (imgIndex == imgList. size())
                 imgIndex = 0 ;
              sfh. unlockCanvasAndPost(c);// 更新屏幕显示内容
           }
        }
     };

     class DrawImage extends Thread {
        int x, y;
        public DrawImage(int x, int y) {
           this . x = x;
           this . y = y;
        }
        public void run() {
           while( true ) {
              if(bitmap！=null ) { //如果图像有效
                 Canvas c = sfh. lockCanvas(new Rect(this . x, this . y, this . x
                     + imgWidth, this . y + imgHeight));
                 c. drawBitmap(bitmap, this . x,    this . y,   new   Paint());
                 sfh. unlockCanvasAndPost(c);// 更新屏幕显示内容
              }
           }
        }
     };

     /*
      * 只负责读取图片的线程
```

```
        */
        class LoadImage extends Thread {
            int imgIndex = 0;
            public void run() {
                while(true) {
                    bitmap = BitmapFactory.decodeResource(getResources(),
                        imgList.get(imgIndex));
                    imgIndex++;
                    if(imgIndex == imgList.size()) //如果到尽头则重新读取
                        imgIndex = 0;
                }
            }
        };
    }
```

事件处理
```
@Override
public void onCreate(Bundle savedInstanceState) {
    super.onCreate(savedInstanceState);

    /* 创建 GameSurfaceView 对象 */
    mGameSurfaceView = new GameSurfaceView(this);

    // 设置显示 GameSurfaceView 视图
    setContentView(mGameSurfaceView);

}
```

习题与思考题

1. SurfaceView 的功能是什么？有何特点？
2. SurfaceView 的刷新为什么要使用多线程？有什么好处？
3. 编写 Android 程序，在 SurfaceView 上实现图片的绘制和刷新，并做出一个小球沿正弦曲线运动的效果。

第 15 章

2D 游戏开发

☆ 15.1 2D游戏开发基础
☆ 15.2 简单游戏框架
☆ 15.3 声音播放
☆ 15.4 手势识别
☆ 15.5 加速度传感器

本章学习目标:掌握 2D 游戏开发的基本知识;能够利用简单游戏框架开发出小游戏;在游戏中加入加速度等传感器元素。

15.1　2D 游戏开发基础

15.1.1　全屏幕显示

代码如下:
```
import android.view.Window;
import android.view.WindowManager;
public class MainActivity extends Activity {
    /** Called when the activity is first created. */
    @Override
    public void onCreate(Bundle savedInstanceState) {
        super.onCreate(savedInstanceState);
        //隐去电池等图标和一切修饰部分(状态栏部分)
        this.getWindow().setFlags(WindowManager.LayoutParams.FLAG_FULLSCREEN,
WindowManager.LayoutParams.FLAG_FULLSCREEN);
        // 隐去标题栏(程序的名字)
        this.requestWindowFeature(Window.FEATURE_NO_TITLE);
        setContentView(new MyView(this));
    }
}
```

15.1.2　画笔无锯齿与背光常亮

代码如下:
```
public class MyView extends View {
    private Paint paint;
    public MyView(Context context) {
        super(context);
        paint = new Paint();
        paint.setAntiAlias(true);//设置画笔无锯齿
        this.setKeepScreenOn(true);//设置背景常亮
        paint.setColor(Color.RED);
    }
    @Override
    public void onDraw(Canvas canvas) {
```

}
}

15.1.3　View 与 SurfaceView

游戏开发的主要内容包括控制逻辑与显示逻辑。Android 中用 View 相关类来完成控制逻辑。

SurfaceView：View 基类中派生出来的显示类，直接子类有 GLSurfaceView 和 VideoView。使用 OpenGL、视频播放以及 Camera 中均使用 SurfaceView。

　　1. 优点

（1）可以控制画面的格式，比如大小、位置。可以通过 Holder 来控制图形控制。

（2）Android 针对 Surface 提供了 GPU 加速功能，已完成实时性很强的图形操作。

（3）surfaceView 是在一个新起的单独线程中可以重新绘制画面，而 View 必须在 UI 的主线程中更新画面。这也是与 View 相比最本质的区别，可以避免过于复杂的操作阻塞 UI 线程，可以使按键、触摸屏等消息能够顺利响应。

　　2. 问题

由于 SurfaceView 是在自己的线程中运行的，这样的话事件也必须传递到 SurfaceView 的线程中处理，复杂时需要有一个时间队列来保存响应的信息，代码复杂度稍高。选用 SurfaceView 和 View 根据下列二种情况而定：

（1）画面被动更新的游戏，比如棋类，这种游戏用 view 就可以。因为画面是依赖于 onTouch 来更新，可以直接使用 invalidate。因为这种情况下，这一次 Touch 和下一次的 Touch 需要的时间比较长些，不会产生影响。

（2）画面主动更新的游戏，比如一个人在一直跑动，这就需要一个单独的 thread 不停重绘人的状态，避免阻塞 main UI thread。所以 view 不合适，需要 SurfaceView 来控制。

Android 中的 SurfaceView 类就是双缓冲机制。因此，开发游戏时尽量使用 SurfaceView 而不要使用 View，这样的话效率较高，而且 SurfaceView 的功能也更加完善。

15.1.4　横竖屏切换

　　1. 指定游戏横竖屏

在 AndroidManifest.xml 里面加入这一行 android：screenOrientation = "landscape"（landscape 是横向，portrait 是纵向）。

　　2. 切换处理

在 android 中每次屏幕的切换动会重启 Activity，所以应该在 Activity 销毁前保存当前活动的状态，在 Activity 再次 Create 的时候载入配置。

　　3. 禁止切换时重启

在 activity 加上 android：configChanges = "keyboardHidden | orientation"属性，就不会重启 activity。而是去调用 onConfigurationChanged(Configuration newConfig)。

　　4. 举例

public void onConfigurationChanged(Configuration newConfig) {

```
        try {
            super.onConfigurationChanged(newConfig);
            if(this.getResources().getConfiguration().orientation == Configuration.
ORIENTATION_LANDSCAPE) {
                Log.v("Himi", "onConfigurationChanged_ORIENTATION_LANDSCAPE");
            } else if(this.getResources().getConfiguration().orientation == Configuration.
ORIENTATION_PORTRAIT) {
                Log.v("Himi", "onConfigurationChanged_ORIENTATION_PORTRAIT");
            }
        } catch (Exception ex) {
        }
    }
```

```xml
<?xml version="1.0" encoding="utf-8"?>
<manifest xmlns:android="http://schemas.android.com/apk/res/android"
    package="com.himi" android:versionCode="1" android:versionName="1.0">
    <application android:icon="@drawable/icon" android:label="@string/app_name">
        <activity android:name=".MainActivity" android:label="@string/app_name"
            android:screenOrientation="landscape" android:configChanges=
"keyboardHidden|orientation">
            <intent-filter>
                <action android:name="android.intent.action.MAIN" />
                <category android:name="android.intent.category.LAUNCHER" />
            </intent-filter>
        </activity>
    </application>
    <uses-sdk android:minSdkVersion="4" />
</manifest>
```

15.2 简单游戏框架

说明：实现对图片的操作及事件处理。代码如下：

```
import android.content.Context;
import android.content.res.Resources;
import android.graphics.Bitmap;
import android.graphics.BitmapFactory;
import android.graphics.Canvas;
import android.graphics.Color;
import android.graphics.Paint;
```

```java
import android.util.Log;
import android.view.KeyEvent;
import android.view.SurfaceHolder;
import android.view.SurfaceView;
import android.view.SurfaceHolder.Callback;
public class MySurfaceView extends SurfaceView implements Callback, Runnable {
    private Thread th = new Thread(this);
    private SurfaceHolder sfh;
    private int SH, SW;
    private Canvas canvas;
    private Paint p;
    private Paint p2;
    private Resources res;
    private Bitmap bmp;
    private int bmp_x = 100, bmp_y = 100;
    private boolean UP, DOWN, LEFT, RIGHT;
    private int animation_up[] = {3, 4, 5};
    private int animation_down[] = {0, 1, 2};
    private int animation_left[] = {6, 7, 8};
    private int animation_right[] = {9, 10, 11};
    private int animation_init[] = animation_down;
    private int frame_count;
    public MySurfaceView(Context context) {
        super(context);
        this.setKeepScreenOn(true);
        res = this.getResources();
        bmp = BitmapFactory.decodeResource(res, R.drawable.enemy1);
        sfh = this.getHolder();
        sfh.addCallback(this);
        p = new Paint();
        p.setColor(Color.YELLOW);
        p2 = new Paint();
        p2.setColor(Color.RED);
        p.setAntiAlias(true);
        setFocusable(true);
    }
    public void draw() {
        canvas = sfh.lockCanvas();
        canvas.drawRect(0, 0, SW, SH, p);
        canvas.save();
```

```
        canvas.drawText("Himi", bmp_x-2, bmp_y-10, p2);
        canvas.clipRect(bmp_x, bmp_y, bmp_x + bmp.getWidth() / 13, bmp_y+bmp.
getHeight());
        if (animation_init == animation_up) {
            canvas.drawBitmap(bmp, bmp_x - animation_up[frame_count] * (bmp.
getWidth() / 13), bmp_y, p);
        } else if (animation_init == animation_down) {
            canvas.drawBitmap(bmp, bmp_x - animation_down[frame_count] * (bmp.
getWidth() / 13), bmp_y, p);
        } else if (animation_init == animation_left) {
            canvas.drawBitmap(bmp, bmp_x - animation_left[frame_count] * (bmp.
getWidth() / 13), bmp_y, p);
        } else if (animation_init == animation_right) {
            canvas.drawBitmap(bmp, bmp_x - animation_right[frame_count] * (bmp.
getWidth() / 13), bmp_y, p);
        }
        canvas.restore();    //备注3
        sfh.unlockCanvasAndPost(canvas);
    }
    public void cycle() {
        if (DOWN) {
            bmp_y += 5;
        } else if (UP) {
            bmp_y -= 5;
        } else if (LEFT) {
            bmp_x -= 5;
        } else if (RIGHT) {
            bmp_x += 5;
        }
        if (DOWN || UP || LEFT || RIGHT) {
            if (frame_count < 2) {
                frame_count++;
            } else {
                frame_count = 0;
            }
        }
        if (DOWN == false && UP == false && LEFT == false && RIGHT ==
false) {
            frame_count = 0;
        }
```

```java
}
@Override
public boolean onKeyDown(int key, KeyEvent event) {
    if (key == KeyEvent.KEYCODE_DPAD_UP) {
        if (UP == false) {
            animation_init = animation_up;
        }
        UP = true;
    } else if (key == KeyEvent.KEYCODE_DPAD_DOWN) {
        if (DOWN == false) {
            animation_init = animation_down;
        }
        DOWN = true;
    } else if (key == KeyEvent.KEYCODE_DPAD_LEFT) {
        if (LEFT == false) {
            animation_init = animation_left;
        }
        LEFT = true;
    } else if (key == KeyEvent.KEYCODE_DPAD_RIGHT) {
        if (RIGHT == false) {
            animation_init = animation_right;
        }
        RIGHT = true;
    }
    return super.onKeyDown(key, event);
}
@Override
public boolean onKeyUp(int keyCode, KeyEvent event) {
    if (DOWN) {
        DOWN = false;
    } else if (UP) {
        UP = false;
    } else if (LEFT) {
        LEFT = false;
    } else if (RIGHT) {
        RIGHT = false;
    }
    return super.onKeyUp(keyCode, event);
}
@Override
```

```
    public void run() {
      // TODO Auto-generated method stub
      while (true) {
        draw();
        cycle();
        try {
          Thread.sleep(100);
        } catch (Exception ex) {
        }
      }
    }
  @Override
  public void surfaceCreated(SurfaceHolder holder) {
      SH = this.getHeight();
      SW = this.getWidth();
      th.start();
  }
  @Override
  public void surfaceChanged(SurfaceHolder holder, int format, int width, int height) {
      // TODO Auto-generated method stub
  }
  @Override
  public void surfaceDestroyed(SurfaceHolder holder) {
      // TODO Auto-generated method stub
  }
}
```

15.3 声音播放

15.3.1 用 MediaPlayer 和 SoundPool 播放声音

游戏开发中的两个主要声音播放方法：MediaPlayer 类与 SoundPool 类。

1. 声音播放

声音播放的代码如下。

```
import java.util.HashMap;
import android.content.Context;
import android.graphics.Canvas;
import android.graphics.Color;
```

```java
import android.graphics.Paint;
import android.media.AudioManager;
import android.media.MediaPlayer;
import android.media.SoundPool;
import android.view.KeyEvent;
import android.view.MotionEvent;
import android.view.SurfaceHolder;
import android.view.SurfaceView;
import android.view.SurfaceHolder.Callback;
    private Thread th;
    private SurfaceHolder sfh;
    private Canvas canvas;
    private Paint paint;
    private boolean ON = true;
    private int currentVol, maxVol;
    private AudioManager am;
    private MediaPlayer player;
    private HashMap<Integer, Integer> soundPoolMap
    private int loadId;
    private SoundPool soundPool;
    public MySurfaceView(Context context) {
        super(context);
// 获取音频服务然后强转成一个音频管理器, 后面方便用来控制音量大小用
        am = (AudioManager) MainActivity.instance
                .getSystemService(Context.AUDIO_SERVICE);
        maxVol = am.getStreamMaxVolume(AudioManager.STREAM_MUSIC);
        // 获取最大音量值
        sfh = this.getHolder();
        sfh.addCallback(this);
        th = new Thread(this);
        this.setKeepScreenOn(true);
        setFocusable(true);
        paint = new Paint();
        paint.setAntiAlias(true);

        //MediaPlayer 的初始化
        player = MediaPlayer.create(context, R.raw.himi);
        player.setLooping(true);//设置循环播放
        //SoundPool 的初始化
        soundPool = new SoundPool(4, AudioManager.STREAM_MUSIC, 100);
```

```java
        soundPoolMap = new HashMap<Integer, Integer>();
        soundPoolMap.put(1, soundPool.load(MainActivity.content,
            R.raw.himi_ogg, 1));
        loadId = soundPool.load(context, R.raw.himi_ogg, 1);
        //load()方法的最后一个参数标识优先考虑的声音。目前没有任何效果。使用了也只是对未来的兼容性价值。
    }
    public void surfaceCreated(SurfaceHolder holder) {
        /*调整媒体音量*/
        MainActivity.instance.setVolumeControlStream(AudioManager.STREAM_MUSIC);

        player.start();
        th.start();
    }
    public void draw() {
        canvas = sfh.lockCanvas();
        canvas.drawColor(Color.WHITE);
        paint.setColor(Color.RED);
        //显示代码
        sfh.unlockCanvasAndPost(canvas);
    }
    private void logic() {
        currentVol = am.getStreamVolume(AudioManager.STREAM_MUSIC);//不断获取当前的音量值
    }
    @Override
    public boolean onKeyDown(int key, KeyEvent event) {
        if (key == KeyEvent.KEYCODE_DPAD_CENTER) {
            ON = !ON;
            if (ON == false)
                player.pause();
            else
                player.start();
        } else if (key == KeyEvent.KEYCODE_DPAD_UP)
            player.seekTo(player.getCurrentPosition() + 5000);
        } else if (key == KeyEvent.KEYCODE_DPAD_DOWN) {
            if (player.getCurrentPosition() < 5000) {
                player.seekTo(0);
            } else {
```

```java
            player.seekTo(player.getCurrentPosition() - 5000);
        }
    }
    else if (key == KeyEvent.KEYCODE_DPAD_LEFT) {
        currentVol += 1;
        if (currentVol > maxVol) {
            currentVol = 100;
        }
        am.setStreamVolume(AudioManager.STREAM_MUSIC, currentVol,
                AudioManager.FLAG_PLAY_SOUND);
    } else if (key == KeyEvent.KEYCODE_DPAD_RIGHT) {
        currentVol -= 1;
        if (currentVol <= 0) {
            currentVol = 0;
        }
        am.setStreamVolume(AudioManager.STREAM_MUSIC, currentVol,
                AudioManager.FLAG_PLAY_SOUND);
    }
    soundPool.play(loadId, currentVol, currentVol, 1, 0, 1f);
    return super.onKeyDown(key, event);
}

@Override
public boolean onTouchEvent(MotionEvent event) {
    return true;
}
public void run() {
    // TODO Auto-generated method stub
    while (true) {
        draw();
        logic();
        try {
            Thread.sleep(100);
        } catch (Exception ex) {
        }
    }
}
public void surfaceChanged(SurfaceHolder holder, int format, int width,
        int height) {
```

```
        }
        public void surfaceDestroyed(SurfaceHolder holder) {
        }
}
```

对代码的一点说明：

HashMap 不同步、空键值、效率高；Hashtable 同步、非空键值、效率略低；这里使用 hashMap 主要是为了存入多个音频的 ID，播放的时候可以同时播放多个音频。

soundPool.play(loadId, currentVol, currentVol, 1, 0, 1f)；

第一个参数指的就是之前的 loadId，是通过 soundPool.load(context, R.raw.himi_ogg, 1)；方法取出来的，返回的音频对应 id，第二个第三个参数表示左右声道大小，第四个参数是优先级，第五个参数是循环次数，最后一个是播放速率(1.0＝正常播放，范围是 0.5 至 2.0)。

注意：必须使用 HashMap<Integer, Integer> hm ＝ new Hash<Integer, Integer>这种形式，否则在使用 SoundPool.get()时会报错。

2. MediaPlayer 播放音频的步骤

(1) 调用 MediaPlayer.create(context, R.raw.himi)；利用 MediaPlayer 类调用 create 方法并且传入通过 id 索引的资源音频文件，得到实例。

(2) 得到的实例就可以调用 MediaPlayer.start()。

3. SoundPool 播放

(1) new 出一个实例；new SoundPool(4, AudioManager.STREAM_MUSIC, 100)；第一个参数是允许有多少个声音流同时播放，第 2 个参数是声音类型，第三个参数是声音的品质。

(2) loadId ＝ soundPool.load(context, R.raw.himi_ogg, 1)。

(3) 使用实例调用 play 方法传入对应的音频文件 id。

4. MediaPlayer 的不足

(1) 资源占用量较高。

(2) 延迟时间较长。

(3) 不支持多个音频同时播放等。

上述缺点决定了 MediaPlayer 在对时间精准度要求相对较高的、有连续快速多个声音播放的游戏开发中不十分适用。

5. SoundPool 播放的不足

(1) SoundPool 最大只能申请 1M 的内存空间，这就意味着我们只能使用一些很短的声音片段，而不是用它来播放歌曲或者游戏背景音乐。

(2) SoundPool 提供了 pause 和 stop 方法，但使用时可能不能够立即停止对声音的播放。

(3) 音频格式建议使用 OGG 格式，对 WAV 的处理稍弱，处理稍慢。

(4) 在初始化中调用播放时，可能会无声音，此时 SoundPool 可能处于准备状态。

15.3.2 音量调整

通过媒体服务得到一个音频管理器，从而来对音量大小进行调整。

调整音频是用这个音频管理器调用 setStreamVolume( ) 的方式去调整，而不是

MediaPlayer.setVolue(int LeftVolume,int RightVolume)，这个方法的两个参数也是调整左右声道而不是调节声音大小。

1. 回放完毕的处理

文件正常播放完毕，而又没有设置循环播放的话就进入 Playback Completed 状态，并会触发 OnCompletionListener 的 onCompletion()方法。此时可以调用 start()方法重新从头播放文件，也可以 stop()停止 MediaPlayer，或者也可以 seekTo()来重新定位播放位置。

绑定：mp.setOnCompletionListener(this);

注意：如果设置了循环播放 mp.setLooping(true);的话，那么永远都不会监听到播放完成的状态。

2. 触摸事件

注意内容：onTouchEvent()中需要根据情况决定是否调用 super.onTouchEvent.，初始化时，需要调用 setFocusableInTouchMode(true)获取触摸模式焦点。

3. 触摸优化

(1) 模拟器的触摸事件

先触发 ACTION_DOWN 然后 ACTION_UP；如果是在屏幕上移动，那么才会触发 ACTION_MOVE 的动作。

(2) 真机

长时间按下即使位置没有变化也会触发 ACTION_MOVE 动作。

(3) 结果

在真机上会大量占用处理时间，造成画面显示延迟。

(4) 改进

```
@Override
public boolean onTouchEvent(MotionEvent event) {
    if (event.getAction() == MotionEvent.ACTION_DOWN) {
        Log.v("Himi", "ACTION_DOWN");
    } else if (event.getAction() == MotionEvent.ACTION_UP) {
        Log.v("Himi", "ACTION_UP");
    } else if (event.getAction() == MotionEvent.ACTION_MOVE) {
        Log.v("Himi", "ACTION_MOVE");
    }
    synchronized (object) {
        try {
            object.wait(50);
        } catch (InterruptedException e) {
            // TODO Auto-generated catch block
            e.printStackTrace();
        }
    }
    return true;//这里一定要返回 true！
}
```

15.4 手势识别

两种手势识别方法：
(1) 触摸屏手势识别。
(2) 输入法手势识别。
类：
android.view.GestureDetector.OnGestureListener；
android.view.GestureDetector；
package com.himi；
import java.util.Vector；
import android.content.Context；
import android.graphics.Bitmap；
import android.graphics.BitmapFactory；
import android.graphics.Canvas；
import android.graphics.Color；
import android.graphics.Paint；
import android.util.Log；
import android.view.GestureDetector；
import android.view.MotionEvent；
import android.view.SurfaceHolder；
import android.view.SurfaceView；
import android.view.View；
import android.view.GestureDetector.OnGestureListener；
import android.view.SurfaceHolder.Callback；
import android.view.View.OnTouchListener；
public class MySurfaceViewAnimation extends SurfaceView implements Callback，
　　Runnable，OnGestureListener，OnTouchListener {
　　private Thread th = new Thread(this)；
　　private SurfaceHolder sfh；
　　private Canvas canvas；
　　private Paint paint；
　　private Bitmap bmp；
　　private GestureDetector gd；
　　private int bmp_x, bmp_y；
　　private boolean isChagePage；
　　private Vector<String> v_str；// 备注1
　　public MySurfaceViewAnimation(Context context) {

```java
        super(context);
        v_str = new Vector<String>();
        this.setKeepScreenOn(true);
        bmp = BitmapFactory.decodeResource(getResources(),
            R.drawable.himi_dream);
        sfh = this.getHolder();
        sfh.addCallback(this);
        paint = new Paint();
        paint.setAntiAlias(true);
        this.setLongClickable(true);
        this.setOnTouchListener(this);// 将本类绑定触屏监听器
        gd = new GestureDetector(this);
        gd.setIsLongpressEnabled(true);
    }
    public void draw() {
        canvas = sfh.lockCanvas();
        if (canvas != null) {
            canvas.drawColor(Color.WHITE);// 画布刷屏
            canvas.drawBitmap(bmp, bmp_x, bmp_y, paint);
            paint.setTextSize(20);// 设置文字大小
            paint.setColor(Color.WHITE);
            //这里画出一个矩形方便童鞋们看到手势操作调用的函数都是哪些
            canvas.drawRect(50, 30, 175, 120, paint);
            paint.setColor(Color.RED);// 设置文字颜色
            if (v_str != null) {
                for (int i = 0; i < v_str.size(); i++) {
                    canvas.drawText(v_str.elementAt(i), 50, 50 + i * 30,
                        paint);
                }
            }
        }
    }
    @Override
    public void run() {
        // TODO Auto-generated method stub
        while (true) {
            draw();
            try {
                Thread.sleep(100);
```

```
            } catch (Exception ex) {
            }
        }
    }
    public void surfaceChanged(SurfaceHolder holder, int format, int width,
        int height) {
    }
    public void surfaceDestroyed(SurfaceHolder holder) {
    }
    @Override
    public boolean onTouch(View v, MotionEvent event) {
        if (v_str ! = null)
            v_str. removeAllElements();
        return gd. onTouchEvent(event);
    }
    @Override
    public boolean onDown(MotionEvent e) {
        // ACTION_DOWN
        v_str. add("onDown");
        return false;
    }
    @Override
    // ACTION_DOWN、短按不移动
    public void onShowPress(MotionEvent e) {
        v_str. add("onShowPress");
    }
    @Override
    // ACTION_DOWN、长按不滑动
    public void onLongPress(MotionEvent e) {
        v_str. add("onLongPress");
    }
    @Override
    // ACTION_DOWN、慢滑动
    public boolean onScroll(MotionEvent e1, MotionEvent e2, float distanceX,
        float distanceY) {
        v_str. add("onScroll");
        return false;
    }
    @Override
    // 短按 ACTION_DOWN、ACTION_UP
```

```java
        public boolean onSingleTapUp(MotionEvent e) {
            v_str.add("onSingleTapUp");
            return false;
        }
    @Override
        // ACTION_DOWN、快滑动、ACTION_UP
        public boolean onFling(MotionEvent e1, MotionEvent e2, float velocityX,
                float velocityY) {
            v_str.add("onFling");
            if (isChagePage)
                bmp = BitmapFactory.decodeResource(getResources(),
                    R.drawable.himi_dream);
            else
                bmp = BitmapFactory.decodeResource(getResources(),
                    R.drawable.himi_warm);
            isChagePage = !isChagePage;
            return false;
        }
```

输入法手势识别的两个重点：

(1) 如何创建输入法手势、删除输入法手势、从 SD 卡中读取出手势文件。

(2) 当输入法手势创建后，如何来匹配出我们的自定义手势。

需要熟悉的内容：

(1) GestureOverlayView：手写绘图区。

(2) GestureLibrary：对手势进行保存、删除等操作。

(3) 笔画类型：单一笔画和多笔画。

```xml
<?xml version="1.0" encoding="utf-8"?>
<LinearLayout xmlns:android="http://schemas.android.com/apk/res/android"
    android:orientation="vertical" android:layout_width="fill_parent"
    android:layout_height="fill_parent">
    <TextView android:id="@+id/himi_tv" android:layout_width="fill_parent"
        android:layout_height="wrap_content" android:text="@string/hello"
        android:textSize="15sp" android:textColor="#FFFFFF00" />
    <EditText android:id="@+id/himi_edit" android:layout_width="fill_parent"
        android:layout_height="wrap_content" />
    <RelativeLayout android:layout_width="fill_parent"
        android:layout_height="wrap_content" android:layout_weight="1">
        <com.himi.MySurfaceView android:id="@+id/view3d"
            android:layout_width="fill_parent" android:layout_height="fill_parent" />
        <android.gesture.GestureOverlayView
```

```xml
            android:id="@+id/himi_gesture" android:layout_width="fill_parent"
            android:layout_height="fill_parent" android:layout_weight="1.0"/>
    </RelativeLayout>
</LinearLayout>
```

```java
public class MainActivity extends Activity {
    private GestureOverlayView gov;// 创建一个手写绘图区
    private Gesture gesture;// 手写实例
    private GestureLibrary gestureLib;//创建一个手势仓库
    private TextView tv;
    private EditText et;
    private String path;//手势文件路径
    private File file;//
    @Override
    public void onCreate(Bundle savedInstanceState) {
        super.onCreate(savedInstanceState);
        this.getWindow().setFlags(WindowManager.LayoutParams.FLAG_FULLSCREEN,
                WindowManager.LayoutParams.FLAG_FULLSCREEN);
        this.requestWindowFeature(Window.FEATURE_NO_TITLE);
        setContentView(R.layout.main);
        tv = (TextView) findViewById(R.id.himi_tv);
        et = (EditText) findViewById(R.id.himi_edit);

        gov = (GestureOverlayView) findViewById(R.id.himi_gesture);
        gov.setGestureStrokeType(GestureOverlayView.GESTURE_STROKE_TYPE_MULTIPLE);//设置笔画类型
        // GestureOverlayView.GESTURE_STROKE_TYPE_MULTIPLE 设置支持多笔画
        // GestureOverlayView.GESTURE_STROKE_TYPE_SINGLE 仅支持单一笔画
        path = new File(Environment.getExternalStorageDirectory(), "gestures").getAbsolutePath();
        //得到默认路径和文件名/sdcard/gestures
        file = new File(path);//实例gestures的文件对象
        gestureLib = GestureLibraries.fromFile(path);//实例手势仓库
        gov.addOnGestureListener(new OnGestureListener() { // 这里是绑定手写绘图区
            @Override
            // 以下方法是你刚开始画手势的时候触发
            public void onGestureStarted(GestureOverlayView overlay, MotionEvent event) {
```

```
                tv.setText("请您在紧凑的时间内用两笔画来完成一个手势!");
            }

            @Override
            // 以下方法是当手势完整形成的时候触发
            public void onGestureEnded(GestureOverlayView overlay, MotionEvent event) {
                gesture = overlay.getGesture();// 从绘图区取出形成的手势
                if (gesture.getStrokesCount() == 2) {//我判定当用户用了两笔画
                    //(强调:如果一开始设置手势笔画类型是单一笔画,那你这里始终得到的
只是1!)
                    if (event.getAction() == MotionEvent.ACTION_UP) {//判定第二笔
画离开屏幕
                        //if(gesture.getLength()==100){}//这里是判定长度达到100像素
                        if (et.getText().toString().equals("")) {
                            tv.setText("保存失败啦～");
                        } else {
                            tv.setText("正在保存手势...");
                            addGesture(et.getText().toString(), gesture);//添加手势函数
                        }
                    }
                } else {
                    tv.setText("请您在紧凑的时间内用两笔画来完成一个手势
                }
            }

            @Override
            public void onGestureCancelled(GestureOverlayView overlay, MotionEvent event) {
            }
            @Override
            public void onGesture(GestureOverlayView overlay, MotionEvent event) {
            }
        });
        //————这里是在程序启动的时候进行遍历所有手势!——————
        if (!gestureLib.load()) {
            tv.setText("");
        } else {
            Set<String> set = gestureLib.getGestureEntries();//取出所有手势
            Object ob[] = set.toArray();
            loadAllGesture(set, ob);
```

```java
            }
        }
    }

    public void addMyGesture(String name, Gesture gesture) {
        try {
            if (name.equals("himi")) {
                findGesture(gesture);
            } else {

                if (Environment.getExternalStorageState() != null) {// 这个方法在
试探终端是否有 sdcard!
                    if (!file.exists()) {// 判定是否已经存在手势文件
                        // 不存在文件的时候我们去直接把我们的手势文件存入
                        gestureLib.addGesture(name, gesture);
                        if (gestureLib.save()) {// //保存到文件中
                            gov.clear(true);//清除笔画

                            tv.setText("保存手势成功!");
                            et.setText("");
                            gestureToImage(gesture);
                        } else {
                            tv.setText("保存手势失败!");
                        }

                    } else {// 当存在此文件的时候我们需要先删除此手势然后把
新的手势放上
                        //读取已经存在的文件,得到文件中的所有手势
                        if (!gestureLib.load()) {//如果读取失败
                            tv.setText("手势文件读取失败!");
                        } else {//读取成功
                            Set<String> set = gestureLib.getGestureEntries();//
取出所有手势

                            Object ob[] = set.toArray();
                            boolean isHavedGesture = false;
                            for (int i = 0; i < ob.length; i++) {//这里是遍历所有
手势的 name
                                if (((String) ob[i]).equals(name)) {//和我们新添
的手势 name 做对比
                                    isHavedGesture = true;
```

```
                    }
                }
                if (isHavedGesture) {//如果此变量为 true 说明有相同 name 的手势
                    //gestureLib.removeGesture(name,gesture);//删除与当前名字相同的手势
                    gestureLib.removeEntry(name);
                    gestureLib.addGesture(name,gesture);
                } else {
                    gestureLib.addGesture(name,gesture);
                }

                if (gestureLib.save()) {
                    gov.clear(true);//清除笔画
                    gestureToImage(gesture);
                    tv.setText("保存手势成功！当前所有手势一共有:"+ob.length+"个");
                    et.setText("");
                } else {
                    tv.setText("保存手势失败!");
                }
                ////--------以下代码是当手势超过 9 个就全部清空操作--------
                if (ob.length>9) {
                  for (int i=0;i<ob.length;i++) {//这里是遍历删除手势
                        gestureLib.removeEntry((String)ob[i]);
                    }
                    gestureLib.save();

                        if (MySurfaceView.vec_bmp!=null) {
                    MySurfaceView.vec_bmp.removeAllElements();//删除放置手势图的容器
                    }
                    tv.setText("手势超过9个,已全部清空!");
                    et.setText("");
                }
                            ob=null;
                set=null;
                }
```

```
                    }
                } else {
                    tv.setText("当前模拟器没有SD卡－－。");
                }
            }
        } catch (Exception e) {
            tv.setText("操作异常!");
        }
    }
```

将手势转化为Bitmap：
```
public void gestureToImage(Gesture ges) {//将手势转换成Bitmap
        //把手势转成图片,存到我们SurfaceView中定义的Image容器中,然后都画出来～
        if (MySurfaceView.vec_bmp != null) {
            MySurfaceView.vec_bmp.addElement(ges.toBitmap(100,100,12,Color.GREEN));
        }
    }
```

遍历手势：
```
public void loadAllGesture(Set<String> set, Object ob[]) { //遍历所有的手势
        if (gestureLib.load()) {//读取最新的手势文件
            set = gestureLib.getGestureEntries();//取出所有手势
            ob = set.toArray();
            for (int i = 0; i < ob.length; i++) {
                //把手势转成Bitmap
                gestureToImage(gestureLib.getGestures((String) ob[i]).get(0));
                //这里是把我们每个手势的名字也保存下来
                MySurfaceView.vec_string.addElement((String) ob[i]);
            }
        }
    }
```

手势匹配：
```
public void findGesture(Gesture gesture) {
    try {
        if (Environment.getExternalStorageState() != null) {//这个方法在试探终端是否有sdcard!
            if (!file.exists()) {// 判定是否已经存在手势文件
                tv.setText("手势文件不存在!!");
            } else {
```

```
                //读取已经存在的文件,得到文件中的所有手势
                if (! gestureLib.load()) {//如果读取失败
                    tv.setText("匹配手势失败,手势文件读取失败!");
                } else {//读取成功
                    List<Prediction> predictions = gestureLib.recognize(gesture);
                    // recognize()的返回结果是一个 prediction 集合,包含了所有与
gesture 相匹配的结果。匹配的结果可能包括多个相似的结果
                    if (! predictions.isEmpty()) {
                        Prediction prediction = predictions.get(0);
                        //prediction 的 score 属性代表了与手势的相似程度 prediction
的 name 代表手势对应的名称 prediction 的 score 属性代表了与 gesture 得相似程度
                        if (prediction.score >= 1) {
                            tv.setText("当前你的手势在手势库中找到最相似的手
势:name =" + prediction.name);
                        }
                    }
                }
            } else {
                tv.setText("匹配手势失败,当前模拟器没有 SD 卡－－。");
            }
        } catch (Exception e) {
            e.printStackTrace();
            tv.setText("由于出现异常,匹配手势失败啦～");
        }
    }
//放在 raw 文件夹下的操作
mLibrary = GestureLibraries.fromRawResource(this, R.raw.gestures);
mLibrary.load();
List<Prediction> predictions = mLibrary.recognize(gesture);
```

15.5 加速度传感器

1. Android 中支持的传感器
(1) 加速度传感器(重力传感器)
(2) 陀螺仪传感器
(3) 光传感器
(4) 恒定磁场传感器
(5) 方向传感器

(6) 恒定的压力传感器
(7) 接近传感器
(8) 温度传感器

```java
public class MySurfaceView extends SurfaceView implements Callback,Runnable{
    private Thread th = new Thread(this);
    private SurfaceHolder sfh;
    private Canvas canvas;
    private Paint paint;
    private SensorManager sm;
    private Sensor sensor;
    private SensorEventListener mySensorListener;
    private int arc_x,arc_y;//圆形的x,y位置
    private float x=0,y=0,z=0;
public MySurfaceView(Context context){
    super(context);
    this.setKeepScreenOn(true);
    sfh = this.getHolder();
    sfh.addCallback(this);
    paint = new Paint();
    paint.setAntiAlias(true);
    setFocusable(true);
    setFocusableInTouchMode(true);
    //通过服务得到传感器管理对象
    sm =(SensorManager)MainActivity.ma.getSystemService(Service.SENSOR_SERVICE);
    sensor = sm.getDefaultSensor(Sensor.TYPE_ACCELEROMETER);//得到一个重力传感器实例
    mySensorListener = new SensorEventListener(){
            @Override
            //传感器获取值发生改变时在响应此函数
            public void onSensorChanged(SensorEvent event){//备注1
                //传感器获取值发生改变,在此处理
                x = event.values[0];//手机横向翻滚
                //x>0 说明当前手机左翻 x<0 右翻
                y = event.values[1];//手机纵向翻滚
                //y>0 说明当前手机下翻 y<0 上翻
                z = event.values[2];//屏幕的朝向
                //z>0 手机屏幕朝上 z<0 手机屏幕朝下
                arc_x -= x;//备注2
```

```
                    arc_y += y;
                }
                @Override
                // 传感器的精度发生改变时响应此函数
                public void onAccuracyChanged(Sensor sensor, int accuracy) {
                    // TODO Auto-generated method stub
                }
            };
        sm.registerListener(mySensorListener, sensor, SensorManager.SENSOR_DELAY_GAME);
        }
        public void surfaceCreated(SurfaceHolder holder) {
            arc_x = this.getWidth() / 2 - 25;
            arc_y = this.getHeight() / 2 - 25;
            th.start();
        }
        public void draw() {
            try {
                canvas = sfh.lockCanvas();
                if (canvas != null) {
                    canvas.drawColor(Color.BLACK);
                    paint.setColor(Color.RED);
                    canvas.drawArc(new RectF(arc_x, arc_y, arc_x + 50,
                        arc_y + 50), 0, 360, true, paint);
                    paint.setColor(Color.YELLOW);
                    canvas.drawText("当前重力传感器的值：", arc_x - 50, arc_y - 30, paint);
                    canvas.drawText("x=" + x + ",y=" + y + ",z=" + z,
                        arc_x - 50, arc_y, paint);
                    String temp_str = "Himi 提示：";
                    String temp_str2 = "";
                    String temp_str3 = "";
                    if (x < 1 && x > -1 && y < 1 && y > -1) {
                        temp_str += "当前手机处于水平放置的状态";
                        if (z > 0) {
                            temp_str2 += "并且屏幕朝上";
                        } else {
                            temp_str2 += "并且屏幕朝下,提示别躺着玩手机,对眼睛不好哟~";
```

```
                    }
                } else {
                    if (x > 1) {
                        temp_str2 += "当前手机处于向左翻的状态";
                    } else if (x < -1) {
                        temp_str2 += "当前手机处于向右翻的状态";
                    }

            if (y > 1) {
                    temp_str2 += "当前手机处于向下翻的状态";
                } else if (y < -1) {
                    temp_str2 += "当前手机处于向上翻的状态";
                }
                if (z > 0) {
                    temp_str3 += "并且屏幕朝上";
                } else {
                    temp_str3 += "并且屏幕朝下,提示别躺着玩手机,对眼睛不好哟~";
                }
            }
                paint.setTextSize(20);
                canvas.drawText(temp_str, 0, 50, paint);
                canvas.drawText(temp_str2, 0, 80, paint);
                canvas.drawText(temp_str3, 0, 110, paint);
            }
        } catch (Exception e) {
            Log.v("Himi", "draw is Error!");
        } finally {
            sfh.unlockCanvasAndPost(canvas);
        }
    }
    @Override
    public void run() {
        // TODO Auto-generated method stub
        while (true) {
            draw();
            try {
                Thread.sleep(100);
            } catch (Exception ex) {
```

```
                }
            }
        }
        public void surfaceChanged(SurfaceHolder holder, int format, int width, int height) {
        }
        public void surfaceDestroyed(SurfaceHolder holder) {
        }
    }
```

2. 传感器速率类型

(1) SENSOR_DELAY_NORMAL   正常

(2) SENSOR_DELAY_UI   适合界面

(3) SENSOR_DELAY_GAME   适合游戏

(4) SENSOR_DELAY_FASTEST   最快

3. 传感器类型

// TYPE_ACCELEROMETER:加速度传感器(重力传感器)类型。

// TYPE_ALL:描述所有类型的传感器。

// TYPE_GYROSCOPE:陀螺仪传感器类型。

// TYPE_LIGHT:光传感器类型。

// TYPE_MAGNETIC_FIELD:恒定磁场传感器类型。

// TYPE_ORIENTATION:方向传感器类型。

// TYPE_PRESSURE:描述一个恒定的压力传感器类型。

// TYPE_PROXIMITY:常量描述型接近传感器。

// TYPE_TEMPERATURE:温度传感器类型描述。

4. 屏幕方向与传感器的关系

(1) 如果当前手机是纵向屏幕:

　　x>0 说明当前手机左翻 x<0 右翻

　　y>0 说明当前手机下翻 y<0 上翻

(2) 如果当前手机是横向屏幕:

　　x>0 说明当前手机下翻 x<0 上翻

　　y>0 说明当前手机右翻 y<0 左翻

5. 传感器的使用步骤

(1) 调用 SensorManager.getDefaultSensor();传入一个想要的传感器的参数得到其实例。

(2) 注册监听器。

(3) 在监听器里处理事件。

6. onSensorChanged 的参数

SensorEvent 对象,包含 Sensor 的最新数据,通过 event.values 获得一个 float[]数组,对于不同的传感器类型,其数组包含的元素个数是不同的,重力传感器总是返回一个长度为 3 的数组,分别代表 X、Y 和 Z 方向的数值。Z 轴表示了手机是屏幕朝上还是屏幕朝下。

习题与思考题

1. SurfaceView 与之前介绍过的 View 对象有什么不同？在游戏开发中一般使用哪一个？

2. 利用本章介绍的简单游戏框架，实现简单的游戏选择菜单界面，可以实现游戏中难度选择等功能的跳转。

3. 在习题 2 的基础上，增加游戏内容，一个小球在屏幕上滚动，滚动的方向受加速度传感器控制。

第 16 章

2D 游戏开发进阶

☆ 16.1　游戏地图
☆ 16.2　碰撞
☆ 16.3　游戏的状态控制
☆ 16.4　打砖块游戏实例

本章学习目标：了解游戏地图的基本概念；掌握碰撞的常用判断方法；掌握游戏状态控制的方法。

16.1 游戏地图

16.1.1 层

层是地图中对于显示内容的描述，通常一张地图是由多个层组成的，比如地表层、建物层、碰撞层等等。如图 16-1 所示。

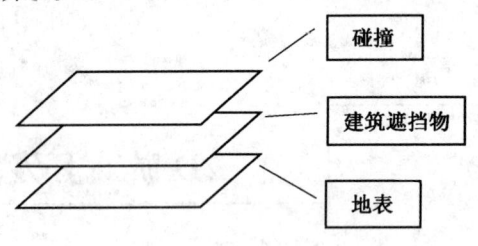

图 16-1 层的内容

16.1.2 Sprite 与动画

Sprite 指精灵，是 2D 游戏中很重要的一个概念。精灵可以是游戏中任意一个可动的对象，如人物、子弹等。

动画指帧动画，是 2D 游戏中最基本的动画显示方法。通常有固定序列动画及复杂切片动画。切片动画需要使用动画编辑器来完成。普通动画只要按照固定的序列来显示图片就可以了。

16.2 碰撞

1. 碰撞分类
(1) 主角与边界的碰撞，限制主角不能走出手机屏幕外。
(2) 主角与物理层的碰撞，与地图中的房子、桌子、椅子等。
(3) 主角与游戏人物之间的碰撞，这里指 NPC 等。
(4) 主角与脚本框发生的碰撞，例如走进房间出现一段剧情对话等等。
可引申为：
① 点与矩形之间的碰撞。
② 矩形与矩形之间的碰撞。
③ 圆形与圆形之间的碰撞。
④ 圆形与矩形之间的碰撞。
2. 地图物理层的碰撞
如图 16-2 所示，在地图的碰撞层中记录了该场景下的碰撞信息，人物运动时检测周边的地图层的碰撞属性,决定是否能够运动。

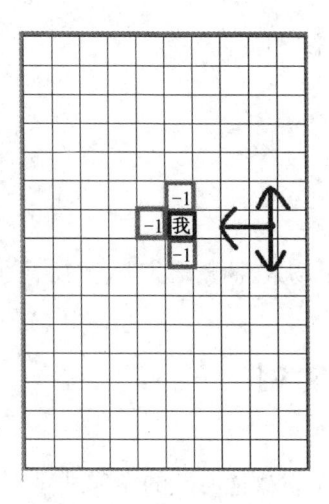

图 16-2　物理层的碰撞

3. 矩形碰撞

如图 16-3 所示,设置碰撞矩形。

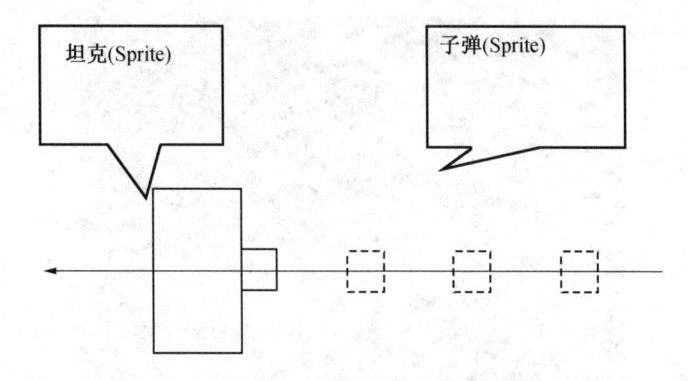

图 16-3　矩形碰撞示图

如图 16-4 所示,一个碰撞实例图。

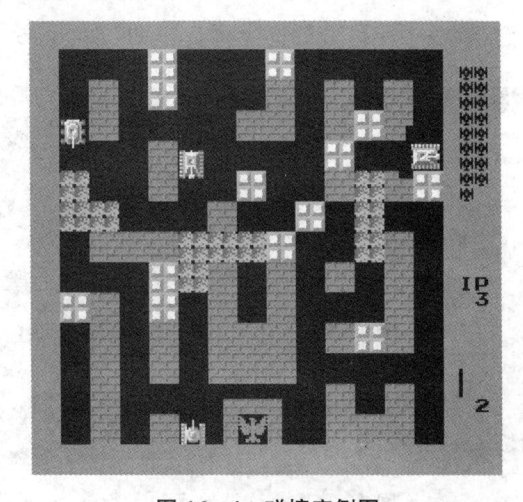

图 16-4　碰撞实例图

16.3 游戏的状态控制

1. 简单的方法
设置一个标志位控制游戏状态。
2. 场景方式
将游戏中的每个画面作为一个场景,抽象出父类,游戏中的各个场景都将是这个类的子类。

16.4 打砖块游戏实例

通过一个打砖块游戏实例来了解碰撞游戏的实现过程。
1. 打砖块游戏描述
初始界面:游戏开始时,游戏上方有 25 个砖,下方有一个挡板(凳子),挡板上有一个小球,初始得分为 0,满分为 100 分。如图 16-5 所示。

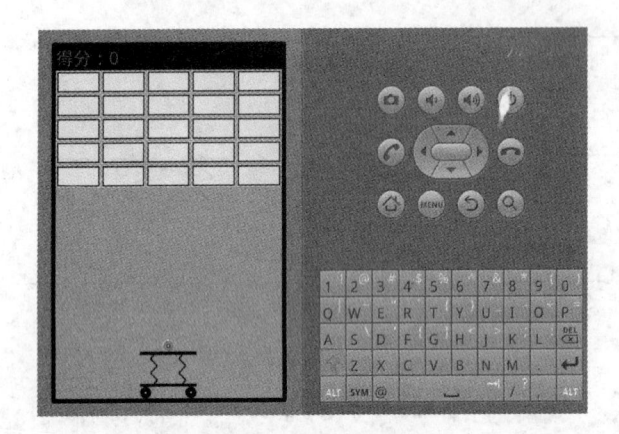

图 16-5 游戏开始图

游戏开始:点击中间键,游戏中小球开始向上弹起运动,如图 16-6 所示。

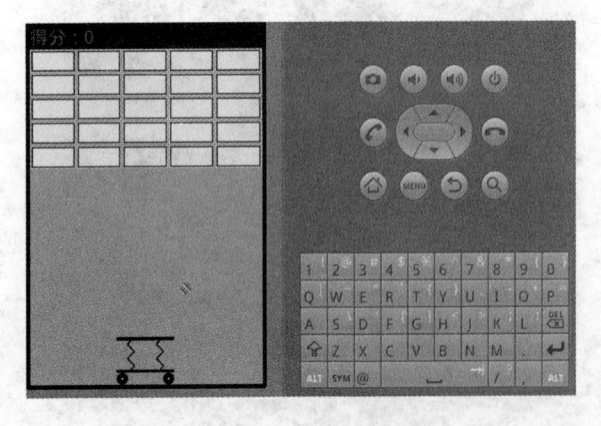

图 16-6 小球开始运动图

第 16 章 2D 游戏开发进阶

游戏计分：当小球弹起运动时，并会敲击游戏屏幕上方的砖块，当小球撞击某一块砖，该砖块会消失，加 4 分，小球并改变运动方向继续运动。当小球向下运动，落下时，如挡板未接到落下的小球，游戏将会扣 10 分，所以要根据小球的运动来移动下面挡板（是方向键来移动），以免被扣分。如图 16-7 所示。

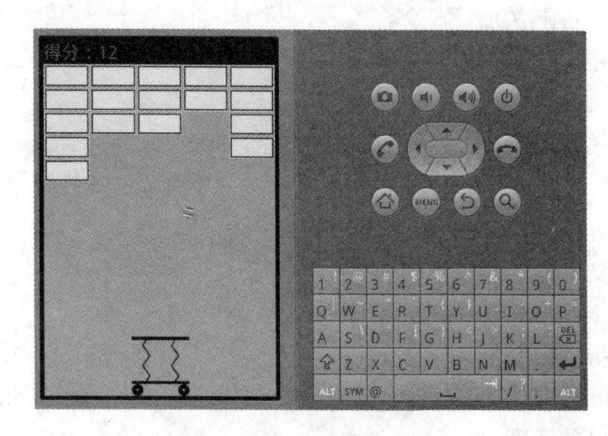

图 16-7　小球运动中的得分

游戏结束：当分数为 0 时或砖块数为 0 游戏结束，显示最终得分，然后，按 MENU 重新开始，按 BACK 结束游戏。如图 16-8 所示。

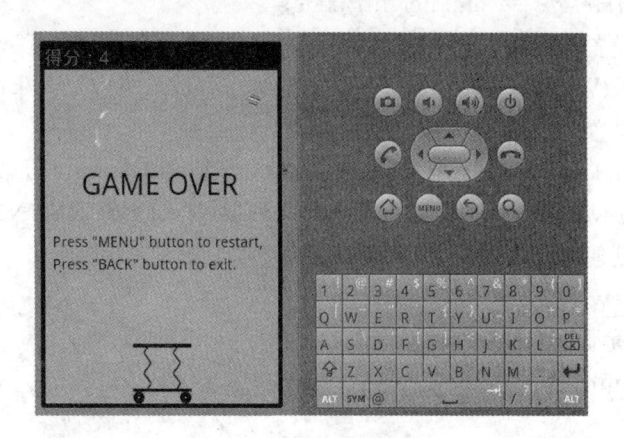

图 16-8　游戏结束

2. 打砖块游戏的程序实现

由于程序不大，主要采用二个类，一个是 ActivityMain 类，一个是 BallView 类。现将代码全部写出，并写有全部注释，如按代码编写，可实现上述游戏。

（1）ActivityMain.java 全部代码和注释如下：

import android.app.Activity;
import android.os.Bundle;
import android.util.DisplayMetrics;
import android.view.KeyEvent;
import android.view.MotionEvent;
import android.view.Window;
import android.view.WindowManager;

```java
public class ActivityMain extends Activity
{
    BallView myView;
    static int screenWidth;
    static int screenHeight;
    @Override
    public void onCreate(Bundle savedInstanceState)
    {
        super.onCreate(savedInstanceState);
        /* 定义 DisplayMetrics 对象 */
        DisplayMetrics dm = new DisplayMetrics();
        /* 取得窗口属性 */
        getWindowManager().getDefaultDisplay().getMetrics(dm);
        /* 窗口的宽度 */
        screenWidth = dm.widthPixels;
        /* 窗口的高度 */
        screenHeight = dm.heightPixels;
        /* 设置为无标题栏 */
        requestWindowFeature(Window.FEATURE_NO_TITLE);
        /* 设置为全屏模式 */
        getWindow().setFlags(WindowManager.LayoutParams.FLAG_FULLSCREEN,
            WindowManager.LayoutParams.FLAG_FULLSCREEN);
        /* 创建 GameSurfaceView 对象 */
        myView = new BallView(this);
        //设置显示 GameSurfaceView 视图
        setContentView(myView);
    }

    //按键事件
    public boolean onKeyDown(int keyCode, KeyEvent event)
    {
    switch (keyCode)
        {
        //菜单按键——设为初始菜单
          case KeyEvent.KEYCODE_MENU:
              myView.resetGame();//重新开始
              break;
        //中间按键
          case KeyEvent.KEYCODE_DPAD_CENTER:
```

```
              myView.ball_isRun = ！myView.ball_isRun;//开始//暂停
              break;
//左方向键
    case KeyEvent.KEYCODE_DPAD_LEFT:
          if(myView.ball_isRun){
            if(myView.board_left<=myView.board_x_move)
              {
                  myView.board_left=0;
                  myView.board_right=myView.board_length;
              }else{
                  myView.board_left-=myView.board_x_move;
                  myView.board_right-=myView.board_x_move;
              }
          }
          break;
//右方向键
    case KeyEvent.KEYCODE_DPAD_RIGHT:
          if(myView.ball_isRun){
              if(screenWidth-myView.board_right<=myView.board_x_move )
              {
                  myView.board_left=screenWidth-myView.board_length;
                  myView.board_right=screenWidth;
              }else{
                  myView.board_left+=myView.board_x_move;
                  myView.board_right+=myView.board_x_move;
              }
          }
          break;
//上方向键
    case KeyEvent.KEYCODE_DPAD_UP:
          if(myView.ball_isRun){
              if( myView.board_alterable_top==myView.board_default_top)
              {
                  myView.board_alterable_top-=myView.boardYadd;
                  myView.board_alterable_bottom-=myView.boardYadd;
              }
          }
          break;
//下方向键
    case KeyEvent.KEYCODE_DPAD_DOWN:
```

```
            if(myView.ball_isRun){
                if(myView.board_alterable_top==myView.board_default_top-myView.boardYadd )
                {
        myView.board_alterable_top=myView.board_default_top;
        myView.board_alterable_bottom=myView.board_alterable_top+myView.board_thickness;
                }
            }
            break;
    //返回键
      case KeyEvent.KEYCODE_BACK:
        this.finish();
        break;
    }
    return false;
}
//按键弹起事件
public boolean onKeyUp(int keyCode, KeyEvent event)
{
    return true;
}
//触笔事件
public boolean onTouchEvent(MotionEvent event)
{
    return true;
}
//按多键事件
public boolean onKeyMultiple(int keyCode, int repeatCount, KeyEvent event)
{
    return true;
}
}
```

（2）BallView.java 全部代码和注释如下：

```
import android.content.Context;
import android.graphics.Canvas;
import android.graphics.Color;
import android.graphics.Paint;
import android.graphics.RadialGradient;
import android.graphics.Shader;
```

```java
import android.util.Log;
import android.view.SurfaceHolder;
import android.view.SurfaceView;
public class BallView extends SurfaceView implements
        SurfaceHolder.Callback, Runnable {
    //线程延时控制
    final int ball_sleep=1;//延时毫秒,延时越大,球速越慢
    final int ball_r= 8;//小球半径
    final float ball2_r= 8;//底下滚珠小球半径
    final int ballXorYadd = 4;//小球的基本位移
    //获取屏幕宽度和高度
    int screen_width;
    int screen_height;
    //砖的属性
    int brick_width;//每块砖宽
    int brick_height;//每块砖高
    boolean brick_exist[];//砖是否存在
    int k ;// 列
    int j ;// 行
    int brick_left = brick_width * (k-1);//到判断语句才初始化
    int brick_right = brick_width * k;
    int brick_top = brick_height * j;
    int brick_bottom = brick_height * (j+1);
    //挡板的属性
    int board_length;//挡板长度
    final int boardYadd = 16;//按上下键时挡板 y 方向位移量
    final int board_x_move = 30;//挡板 x 方向位移量:可以随意自定义
    int board_left;//挡板左侧(可变)
    int board_right;//挡板右侧(可变)
    int board_thickness;//挡板厚度
    int board_default_top;// 即 435,挡板的 top 面初始位置
    int board_alterable_top;//挡板上侧(可变)
    int board_alterable_bottom;//挡板下侧(可变)
    int ball_default_x;//球的初始 x 坐标
    int ball_default_y;//球的初始 y 坐标
    //球的即时坐标(可变):
    int ball_x;//球心横坐标
    int ball_y;//球心纵坐标
    //球的前一步的 y 坐标
    int ball_previous_y;
```

```java
int ball_x_speed;//球的横向偏移量//可变
int ball_y_speed;//球的纵向偏移量//可变
boolean ball_isRun;//球是否在动
// 控制循环
boolean mbLoop;
// 定义 SurfaceHolder 对象
SurfaceHolder mSurfaceHolder = null;
//获得分数
int score;
/* 唤醒渐变渲染 */
Shader mRadialGradient = null;
public BallView(Context context) {
    super(context);
    // 实例化 SurfaceHolder
    mSurfaceHolder = this.getHolder();
    // 添加回调
    mSurfaceHolder.addCallback(this);
    this.setFocusable(true);
    //获取屏幕宽度和高度
    screen_width = ActivityMain.screenWidth;
    screen_height = ActivityMain.screenHeight;
    //砖的属性
    brick_width = screen_width/5;//每块砖宽 64
    brick_height = screen_height/15;//每块砖高 32
    //挡板的属性
    board_length = screen_width/4;//挡板长度:80 比较合适,可以随意修改,但别超过 screen_width
    board_left = (screen_width-board_length)/2;//挡板左侧(可变)
    board_right = (screen_width+board_length)/2;//挡板右侧(可变)
    board_thickness = 5;//挡板厚度
    board_default_top = 13 * screen_height/15;// 即 435,挡板的 top 面初始位置
    board_alterable_top = board_default_top;//挡板上侧(可变)
    board_alterable_bottom = board_alterable_top+board_thickness;//挡板下侧(可变)
    ball_default_x = screen_width/2;//球的初始 x 坐标
    ball_default_y = board_default_top - ball_r;//球的初始 y 坐标
    //球的即时坐标(可变):
    ball_x = ball_default_x;
    ball_y = ball_default_y;
```

```java
            //球的前一步的y坐标
            ball_previous_y = 0;
            ball_x_speed = ballXorYadd;//球的横向偏移量
            ball_y_speed = ballXorYadd;//球的纵向偏移量
            mbLoop = true;
            ball_isRun = false;
            score=0;
            brick_exist = new boolean[25];
            for (int i = 0; i < 25; i++) {
                brick_exist[i] = true;
            }
            /* 构建 RadialGradient 对象,设置半径的属性 */
            mRadialGradient = new RadialGradient(ball_x, ball_y, ball_r,//球中心坐标x,y,半径r
                    new
    int[]{Color.WHITE,Color.BLUE,Color.GREEN,Color.RED,Color.YELLOW},//颜色数组
                    null,//颜色数组中每一种颜色对应的相对位置,为空的话就是平均分布,由中心向外排布
                    Shader.TileMode.REPEAT);//渲染模式
        }
        public void resetGame(){
            ball_isRun = false;
            score =0;//分数
            ball_x_speed = ballXorYadd;//球的横向偏移量
            ball_y_speed = ballXorYadd;//球的纵向偏移量
            ball_x = screen_width/2;//球心起始横坐标
            ball_y = board_default_top - ball_r;//球心起始纵坐标
            board_left = (screen_width-board_length)/2;//挡板左侧
            board_right = (screen_width+board_length)/2;//挡板右侧
            board_alterable_top = board_default_top;//挡板上侧
            board_alterable_bottom = board_alterable_top+board_thickness;
        for (int i = 0; i < 25; i++) {
                brick_exist[i] = true;
            }
        }
        //——————————————————————————绘图循环开始
        public void run() {

            while (mbLoop&&! Thread.currentThread().isInterrupted()){
```

```
            try
            {
                Thread.sleep(ball_sleep);
            }
            catch (InterruptedException e)
            {
                Thread.currentThread().interrupt();
            }
            //球的前一步 y 坐标
            ball_previous_y = ball_y;
            if (ball_isRun)
            {
                ballRunning();// 让小球移动
                boardPositionCheck();//检测挡板是否处于"中线"位置,是就随小球
```
上升一步,直至到"上线"
```
                hitWallCheck();//墙壁碰撞检测
                hitBoardCheck();//挡板碰撞检测
                hitBrickCheck();//砖块碰撞检测
            }
            synchronized (mSurfaceHolder)
            {
                Draw();
            }
        }
    }
    //让小球移动
    public void ballRunning() {
        ball_x += ball_x_speed;
        ball_y -= ball_y_speed;
    }
    //朝左或朝右碰撞后小球水平方向逆向
    public void ballLeftOrRightHit() {
        ball_x_speed *= -1;
    }
    //朝上或朝下碰撞后小球竖直方向逆向
    public void ballUpOrDownHit() {
        ball_y_speed *= -1;
    }
    //碰撞角落
    public void ballcornerHit() {
```

```
            ball_x_speed *= -1;
            ball_y_speed *= -1;
        }
        //——————————————————墙壁碰撞检测开始—
        public void hitWallCheck() {
            // 左碰墙
            if (ball_x <= ball_r &&
                ball_y >= ball_r && ball_y <= screen_height)
            {
                ballLeftOrRightHit();
            }
            // 右碰墙
            if (ball_x >= screen_width - ball_r &&
                ball_y >= ball_r && ball_y <= screen_height)
            {
                ballLeftOrRightHit();
            }
            // 上碰墙
            if (ball_x >= ball_r && ball_x <= screen_width - ball_r &&
                    ball_y <= ball_r + brick_height)
            {
                ballUpOrDownHit();
            }
            // 下碰墙
            if (ball_x >= ball_r && ball_x <= screen_width - ball_r &&
                ball_y >= screen_height - ball_r)
            {
                ballUpOrDownHit();
                score -= 10;
            }
        }

        //————————————————————挡板碰撞检测开始
        public void hitBoardCheck() {
            // 下碰挡板正面
            if (ball_x >= ball_r && ball_x <= screen_width - ball_r &&  //在屏幕内,起码条件
                ball_x >= board_left && ball_x <= board_right &&  //在挡板 X 域上方
                    ball_y == board_alterable_top - ball_r &&  //球面与挡板相切
                    ball_previous_y <= board_alterable_top - ball_r   //确定球是从上方
```

下落
)
{
　　if(board_alterable_top==board_default_top-boardYadd){//如果弹簧伸张,挡板位于上线
　　　　ballHitBoardlower();//作用:ball_y_move 减小;挡板被打下;小球 Y 向运动反向
　　}
　　else if(board_alterable_top==board_default_top){//如果弹簧压缩,挡板位于下线
　　　　boardHitBallHigher();//作用:ball_y_move 增加;挡板弹上;小球 Y 向运动反向
　　}
}
//斜碰挡板右上角
else　if(Math.pow(board_right-ball_x,2)+Math.pow(board_alterable_top-ball_y,2)<=Math.pow(ball_r,2)&&
　　ball_x>board_right && ball_y<board_alterable_top)
{
　　ballcornerHit();
}
//斜碰挡板的左上角
else　if(Math.pow(board_left-ball_x,2)+Math.pow(board_alterable_top-ball_y,2)<=Math.pow(ball_r,2)&&
　　ball_x<board_left && ball_y<board_alterable_top)
{
　　ballcornerHit();
}
}
//————————————————————————挡板碰撞检测结束
private void boardHitBallHigher(){//增强
　　ballUpOrDownHit();//小球 Y 方向反向,ball_y_speed 变为正数
　　if(ball_y_speed == ballXorYadd){
　　　　ball_y_speed += ballXorYadd;//离开挡板后小球 Y 方向速度增强
　　}
　　if(boardYadd > ball_y_speed){//在线程这一轮,小球上升多少,挡板就上升多少。
　　　　board_alterable_top = board_default_top - ball_y_speed;
　　　　board_alterable_bottom= board_alterable_top+board_thickness;//挡板下层面

            }
        }
        //检测挡板是否处于"中线"位置,是就随小球上升一步,直至到"上线"
        private void boardPositionCheck(){//还可直接利用球的位置刷新,board_top 与球心相差 ball_r
            if(board_alterable_top < board_default_top && board_alterable_top > board_default_top-boardYadd){
                //挡板随球上升
                if(board_alterable_top - ball_y_speed >= board_default_top-boardYadd){
                    board_alterable_top -= ball_y_speed;//挡板上层面
                    board_alterable_bottom= board_alterable_top+board_thickness;//挡板下层面
                }
                else{
                    board_alterable_top = board_default_top-boardYadd;//挡板上层面
                    board_alterable_bottom= board_alterable_top+board_thickness;//挡板下层面
                }
            }
        }

        private void ballHitBoardlower(){//减弱
            board_alterable_top=board_default_top;
            board_alterable_bottom=board_default_top+board_thickness;//挡板被打退
            ballUpOrDownHit();//小球 Y 方向反向
            if(ball_y_speed==2*ballXorYadd){
                ball_y_speed -= ballXorYadd;//小球 Y 方向速度减弱
            }
        }
        //砖块碰撞检测开始
        public void hitBrickCheck() {
            for (int i = 0; i<25 ; i++) {
                if (brick_exist[i])
                {
                    k = i % 5+1;
                    j = i / 5+1;
                    brick_left = brick_width * (k-1);
                    brick_right = brick_width * k;
                    brick_top = brick_height * j;

```
        brick_bottom = brick_height * (j+1);
        //朝下碰砖的 top 面
        if(ball_x >= brick_left && ball_x <= brick_right &&
          ball_y >= brick_top-ball_r && ball_y < brick_top)
        {
            ballUpOrDownHit();
            brick_exist[i] = false;
            score+=4;
            //朝下正碰2砖中间,i砖右上角检测
            if(k! =5 && ball_x == brick_right)//如果不是第5列砖的右
侧边
            {
                //如果砖[i+1]存在
                if(brick_exist[i+1]){
                    brick_exist[i+1] = false;
                    score+=4;
                }
            }
            //朝下正碰2砖中间,i砖左上角检测
            else if(k! =1 && ball_x == brick_left)
            {
                //如果砖[i-1]存在
                if(brick_exist[i-1]){
                    brick_exist[i-1] = false;
                    score+=4;
                }
            }
        }
        //朝上碰砖的 bottom 面
        else if(ball_x >= brick_left && ball_x <= brick_right &&
          ball_y > brick_bottom && ball_y <= brick_bottom + ball_r )
        {
            ballUpOrDownHit();
            brick_exist[i] = false;
            score+=4;
            //朝上正碰2块砖中间——i砖的右下角检测
            if(k! =5 && ball_x == brick_right) //如果不是第5列砖的右侧边
            {
                if(brick_exist[i+1]){//如果砖[i+1]存在
                    brick_exist[i+1] = false;
```

```
            score+=4;
        }
    }
    //朝上正碰2块砖中间——i砖的左下角检测
    else if(k!=1 && ball_x == brick_left)　//如果不是第1列砖的左侧边
    {
        if(brick_exist[i-1]){//如果砖[i-1]存在
            brick_exist[i-1] = false;
            score+=4;
        }
    }
}
//朝右碰砖的left面
else if(ball_x >= brick_left-ball_r && ball_x < brick_left &&
    ball_y >= brick_top && ball_y <= brick_bottom)
{
    ballLeftOrRightHit();
    brick_exist[i] = false;
    score+=4;
    //朝右正碰2块砖中间,左下角检测
    if(j!=5 && ball_y == brick_bottom)//如果不是第5行砖的下侧边
    {
      if(brick_exist[i+5]){//如果砖[i+5]存在
            brick_exist[i+5] = false;
            score+=4;
        }
    }
    //朝右正碰2块砖中间,左上角检测
    else if(j!=1 && ball_y == brick_top)//如果不是第1行砖的上侧边
    {
        if(brick_exist[i-5]){//如果砖[i-5]存在
            brick_exist[i-5] = false;
            score+=4;
        }
    }
}
//朝左碰砖的right面
else if(ball_x >= brick_right && ball_x <= brick_right+ball_r &&
    ball_y >= brick_top && ball_y <= brick_bottom)
{
```

```
            ballLeftOrRightHit();
            brick_exist[i] = false;
            score+=4;
            //朝左正碰2块砖中间,右下角检测
            if(j!=5 && ball_y == brick_bottom )//如果不是第5行砖的下侧边
            {
                if(brick_exist[i+5]){//如果砖[i+5]存在
                    brick_exist[i+5] = false;
                    score+=4;
                }
            }
            //朝左正碰2块砖中间,右上角检测
            else if(j!=1 && ball_y == brick_top )//如果不是第1行砖上侧边
            {
                if(brick_exist[i-5]){//如果砖[i-5]存在
                    brick_exist[i-5] = false;
                    score+=4;
                }
            }
        }
        //斜碰i砖的左下角
        else if(( i-1<0||(i-1>=0&&! brick_exist[i-1]) )&&
                (i+5>=25||(i+5<25&&! brick_exist[i+5]) )&&
                Math.pow(brick_left-ball_x, 2)+Math.pow(brick_bottom-ball_y,2)
<=Math.pow(ball_r, 2)&&
                ball_x>brick_left-ball_r && ball_x<brick_left &&
                ball_y>brick_bottom && ball_y<brick_bottom+ball_r )
        {
            ballcornerHit();
            brick_exist[i] = false;
            score+=4;
        }
        //斜碰i砖的右下角
        else if( (i+1>=25||(i+1<25&&! brick_exist[i+1]) )&&
                (i+5>=25||(i+5<25&&! brick_exist[i+5]) )&&
                Math.pow(brick_right-ball_x, 2)+Math.pow(brick_bottom-ball_y,
2)<=Math.pow(ball_r, 2)&&
                ball_x>brick_right&&ball_x<brick_right+ball_r&&
                ball_y>brick_bottom&&ball_y<brick_bottom+ball_r )
        {
```

```
            Log.v("----------","right bottom hit"+i+":"+brick_exist[i]);
                ballcornerHit();
                brick_exist[i] = false;
                score+=4;
            }
            //斜碰i砖的右上角
            else if( (i+1>=25||(i+1<25&&! brick_exist[i+1]) )&&
                    (i-5<0||(i-5>0&&! brick_exist[i-5]) )&&
                    Math.pow(brick_right-ball_x, 2)+Math.pow(brick_top-ball_y,2)<=Math.pow(ball_r, 2)&&
                    ball_x>brick_right && ball_x<brick_right+ball_r&&
                    ball_y>brick_top-ball_r && ball_y<brick_top)
            {
                ballcornerHit();
                brick_exist[i] = false;
                score+=4;
            }
            //斜碰i砖的左上角
            else if((i-1<0||(i-1>=0&&! brick_exist[i-1])) &&
                    (i-5<0||(i-5>=0&&! brick_exist[i-5])) &&
                    Math.pow(brick_left-ball_x, 2)+Math.pow(brick_top-ball_y,2)<=Math.pow(ball_r, 2)&&
                    ball_x>brick_left-ball_x && ball_x<brick_left &&
                    ball_y>brick_top-ball_r && ball_y<brick_top )
            {
                ballcornerHit();
                brick_exist[i] = false;
                score+=4;
            }
        }
    }
}
//---------------------------------------
public boolean gameOver(){
    int count = 0;
    for(boolean s:brick_exist){
        if(! s){
            count++;
        }
```

```
        }
        if (score<0)
        {
            return true;
        }
        if(count == 25)
        {
            return true;
        }else{
            return false;
        }
    }
//——————————绘图方法开始——————————
public void Draw() {
    // 锁定画布,得到 canvas
    Canvas canvas = mSurfaceHolder.lockCanvas();
    if (mSurfaceHolder == null || canvas == null) {
        return;
    }
    // 绘图
    Paint mPaint = new Paint();
    // 设置取消锯齿效果
    mPaint.setAntiAlias(true);
    mPaint.setColor(Color.BLACK);
    // 绘制矩形——背景
    canvas.drawRect(0, 0, screen_width, brick_height-2, mPaint);
    mPaint.setColor(Color.GREEN);
    // 绘制矩形——背景
    canvas.drawRect(0, brick_height-2, screen_width, screen_height, mPaint);
    mPaint.setColor(Color.RED);// 设置字体颜色
    mPaint.setTextSize(brick_height-7);// 设置字体大小
    canvas.drawText("得分:"+score, 0, brick_height-7, mPaint);
    // 绘制顶层挡板
    mPaint.setColor(Color.BLACK);// 设置颜色
    mPaint.setStrokeWidth(4);// 设置粗细
    canvas.drawLine(0, brick_height-2, screen_width, brick_height-2, mPaint);
    for (int i = 0; i<25 ; i++)
    {
        if (brick_exist[i])
        {
```

```
            k = i % 5+1;// 1,2,3,4,5 循环
            j = i / 5+1;// 1,1,1,1,1;2,2,2,2,2,;...;5,5,5,5,5
            brick_left = brick_width * (k-1);
            brick_right = brick_width * k;
            brick_top = brick_height * j;
            brick_bottom = brick_height * (j+1);
            mPaint.setStyle(Paint.Style.FILL);// 设置实心画笔
            mPaint.setColor(Color.YELLOW);
            canvas.drawRect(brick_left+1, brick_top+1, brick_right-1,brick_bottom-
1, mPaint);
            mPaint.setStyle(Paint.Style.STROKE);// 设置空心画笔
            mPaint.setStrokeWidth(2);//设置粗细
            mPaint.setColor(Color.BLUE);
            canvas.drawRect(brick_left+3, brick_top+3, brick_right-3,brick_bottom-
3, mPaint);
        }
    }
    // 设置实心画笔
    mPaint.setStyle(Paint.Style.FILL);
    {
        mPaint.setShader(mRadialGradient);
        canvas.drawCircle(ball_x, ball_y, ball_r, mPaint);
    }
    Paint mPaint2 = new Paint();
        // 设置取消锯齿效果
    mPaint2.setAntiAlias(true);
    // 设置实心画笔
    mPaint2.setStyle(Paint.Style.FILL);
    {
    mPaint2.setColor(Color.BLACK);
    /* 绘制矩形挡板 */
    canvas.drawRect(board_left, board_alterable_top, board_right, board_alterable_
bottom,mPaint2);
        float board2_bottom = screen_height - 2 * ball2_r;//164
        float board2_top = board2_bottom - 4;//160
        //线段端点坐标数组
        float x0 = board_left+(board_right-board_left)/4;
        float y0 = board_alterable_bottom;//440 或者 444
        float springAdd = (board2_top- board_alterable_bottom)/8;
        float springWidth = 5.0f;//弹簧小线段的 x 向长度
```

```
float x1 = x0 - springWidth;
float y1 = y0 + springAdd;
float x2 = x0 + springWidth;
float y2 = y0 + 3 * springAdd;
float x3 = x1;
float y3 = y0 + 5 * springAdd;
float x4 = x2;
float y4 = y0 + 7 * springAdd;
float x5 = x0;
float y5 = board2_top;
float between_spring = (board_right - board_left)/2;
float pts[] = {x0,y0,x1,y1,
               x1,y1,x2,y2,
               x2,y2,x3,y3,
               x3,y3,x4,y4,
               x4,y4,x5,y5};
float pts2[] ={x0+between_spring,y0, x1+between_spring,y1,
               x1+between_spring,y1, x2+between_spring,y2,
               x2+between_spring,y2, x3+between_spring,y3,
               x3+between_spring,y3, x4+between_spring,y4,
               x4+between_spring,y4, x5+between_spring,y5};
mPaint2.setStrokeWidth(2);//设置弹簧粗细
//绘制2个弹簧
canvas.drawLines(pts, mPaint2);
canvas.drawLines(pts2, mPaint2);
//绘制下层挡板
canvas.drawRect(board_left, board2_top, board_right, board2_bottom,mPaint2);
mPaint2.setColor(Color.BLACK);
// 绘制最下面的两个"轮子"(圆心 x,圆心 y,半径 r,p)
 canvas.drawCircle(board_left + ball2_r, screen_height - ball2_r, ball2_r, mPaint2);//圆
 canvas.drawCircle(board_right - ball2_r, screen_height - ball2_r, ball2_r, mPaint2);
mPaint2.setColor(Color.WHITE);
canvas.drawPoint(board_left+ball2_r, screen_height-ball2_r, mPaint2);//绘制左轮轮心
canvas.drawPoint(board_right-ball2_r, screen_height-ball2_r, mPaint2);//绘制右轮轮心

}//实心画笔 mPaint2 结束
```

```java
        if(gameOver())
        {
            ball_isRun = false;
            mPaint2.setColor(Color.BLACK);//设置字体颜色
            mPaint2.setTextSize(40.0f);//设置字体大小
            canvas.drawText("GAME OVER", screen_width/32+40, screen_height/16
*9-70, mPaint2);
            mPaint2.setTextSize(20.0f);//设置字体大小
            canvas.drawText("Press \"MENU\" button to restart,",
                    screen_width/32,
                    screen_height/16*9,
                    mPaint2);
            canvas.drawText("Press \"BACK\" button to exit. ",
                screen_width/32,
                screen_height/16*10,
                mPaint2);
        }
        // 绘制后解锁,绘制后必须解锁才能显示
        mSurfaceHolder.unlockCanvasAndPost(canvas);
    }

    // 在 surface 的大小发生改变时激发
    public void surfaceChanged(SurfaceHolder holder, int format, int width,
            int height) {
    }
    // 在 surface 创建时激发
    public void surfaceCreated(SurfaceHolder holder)
    {
        // 开启绘图线程
        new Thread(this).start();
    }
    // 在 surface 销毁时激发
    public void surfaceDestroyed(SurfaceHolder holder) {
        // 停止循环
        mbLoop = false;
    }
}
```

习题与思考题

1. 简述游戏开发中地图的概念，以及使用地图的好处。
2. 简述碰撞的几种常用判断方法。
3. 按本章源提供的码来实现打砖块游戏。

第 17 章

位置服务与地图

☆ 17.1 位置服务
☆ 17.2 Google 地图应用

本章学习目标：了解位置服务的概念；了解地图密钥的申请方法；掌握获取位置信息的方法；掌握 MapView 和 MapController 的使用方法；掌握 Google 地图覆盖层的使用方法。

17.1 位置服务

位置服务（Location-Based Services，LBS），又称定位服务或基于位置的服务，融合了 GPS 定位、移动通信、导航等多种技术，为用户提供了与空间位置相关的综合应用服务。

位置服务首先在日本得到商业化的应用，2001 年 7 月，DoCoMo 发布了第一款具有三角定位功能的手持设备；2001 年 12 月，KDDI 发布第一款具有 GPS 功能的手机。基于位置的服务发展迅速，已涉及商务、医疗、工作和生活的各个方面，为用户提供定位、追踪和敏感区域警告等一系列服务。

Android 平台支持提供位置服务的 API，在开发过程中主要用 LocationManager 和 LocationProviders 对象。LocationManager 可以用来获取当前的位置，追踪设备的移动路线，或设定敏感区域，在进入或离开敏感区域时设备会发出特定警报。LocationProviders 是能够提供定位功能的组件集合，集合中的每种组件以不同的技术提供设备的当前位置，区别在于定位的精度、速度和成本等方面。

提供位置服务，首先需要获得 LocationManager 对象。获取 LocationManager 可以通过调用 android.app.Activity.getSystemService()函数实现，代码如下。

```
1. String serviceString = Context.LOCATION_SERVICE;
2. LocationManager LocationManager = (LocationManager)getSystemService
   (serviceString);
```

代码第 1 行的 Context.LOCATION_SERVICE 指明获取的服务是位置服务。代码第 2 行的 getSystemService()函数，可以根据服务名称获取 Android 提供的系统级服务。

Android 支持的系统级服务表，如表 17-1 所示。

表 17-1 系统级服务表

Context 类的静态常量	值	返回对象	说　明
LOCATION_SERVICE	location	LocationManager	控制位置等设备的更新
WINDOW_SERVICE	window	WindowManager	最顶层的窗口管理器
LAYOUT_INFLATER_SERVICE	layout_inflater	LayoutInflater	将 XML 资源实例化为 View
POWER_SERVICE	power	PowerManager	电源管理
ALARM_SERVICE	alarm	AlarmManager	在指定时间接受 Intent
NOTIFICATION_SERVICE	notification	NotificationManager	后台事件通知
KEYGUARD_SERVICE	keyguard	KeyguardManager	锁定或解锁键盘
SEARCH_SERVICE	search	SearchManager	访问系统的搜索服务

续表

Context 类的静态常量	值	返回对象	说　明
VIBRATOR_SERVICE	vibrator	Vibrator	访问支持振动的硬件
CONNECTIVITY_SERVICE	connection	ConnectivityManager	网络连接管理
WIFI_SERVICE	wifi	WifiManager	Wi-Fi 连接管理
INPUT_METHOD_SERVICE	input_method	InputMethodManager	输入法管理

在获取到 LocationManager 后,还需要指定 LocationManager 的定位方法,然后才能够调用 LocationManager。

LocationManager 支持的定位方法有两种:

(1) GPS 定位:可以提供更加精确的位置信息,但定位速度和质量受到卫星数量和环境情况的影响。

(2) 网络定位:提供的位置信息精度差,但速度较 GPS 定位快。

在指定 LocationManager 的定位方法后,则可以调用 getLastKnowLocation()方法获取当前的位置信息。

LocationManager 支持的定位方法,如表 17-2 所示。

表 17-2　定位参数表

LocationManager 类的静态常量	值	说　明
GPS_PROVIDER	gps	使用 GPS 定位,利用卫星提供精确的位置信息,需要 android. permissions. ACCESS_FINE_LOCATION 用户权限
NETWORK_PROVIDER	network	使用网络定位,利用基站或 Wi-Fi 提供近似的位置信息,需要具有如下权限: android. permission. ACCESS_COARSE_LOCATION 或 android. permission. ACCESS_FINE_LOCATION

以使用 GPS 定位为例,获取位置信息的代码如下:

```
1. String provider = LocationManager.GPS_PROVIDER;
2. Location location = locationManager.getLastKnownLocation(provider);
```

代码第 2 行返回的 Location 对象中,包含了可以确定位置的信息,如经度、纬度和速度等。

通过调用 Location 中的 getLatitude()和 getLonggitude()方法可以分别获取位置信息中的纬度和经度,示例代码如下:

```
1. double lat = location.getLatitude();
2. double lng = location.getLongitude();
```

LocationManager 提供了一种便捷、高效的位置监视方法。requestLocationUpdates()可以根据位置的距离变化和时间间隔设定产生位置改变事件的条件,这样可以避免因微小的距离变化而产生大量的位置改变事件。

LocationManager 中设定监听位置变化的代码如下：

```
locationManager.requestLocationUpdates(provider, 2000, 10, locationListener);
```

第1个参数是定位的方法，GPS定位或网络定位。第2个参数是产生位置改变事件的时间间隔，单位为微秒。第3个参数是距离条件，单位是米。第4个参数是回调函数，在满足条件后的位置改变事件的处理函数。代码将产生位置改变事件的条件设定为距离改变10米，时间间隔为2秒。

实现 locationListener 的代码如下：

```
1.  LocationListener locationListener = new LocationListener(){
2.  public void onLocationChanged(Location location) {
3.  }
4.  public void onProviderDisabled(String provider) {
5.  }
6.  public void onProviderEnabled(String provider) {
7.  }
8.  public void onStatusChanged(String provider, int status, Bundle extras) {
9.  }
10. };
```

第2行代码 onLocationChanged() 在设备的位置改变时被调用。第4行的 onProviderDisabled() 在用户禁用具有定位功能的硬件时被调用。第6行的 onProviderEnabled() 在用户启用具有定位功能的硬件时被调用。第8行的 onStatusChanged() 在提供定位功能的硬件的状态改变时被调用，如从不可获取位置信息状态到可以获取位置信息的状态，反之亦然。

为了使 GPS 定位功能生效，还需要在 AndroidManifest.xml 文件中加入用户许可。实现代码如下：

```
<uses-permission
android:name="android.permission.ACCESS_FINE_LOCATION"/>
```

CurrentLocationDemo 是一个提供位置服务的基本示例，提供了显示当前新位置的功能，并能够监视设备的位置变化，如图 17-1 所示。

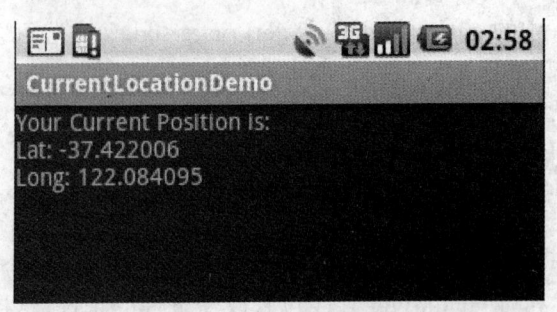

图 17-1 CurrentLocationDemo 示例

第 17 章 位置服务与地图

位置服务一般都需要使用设备上的硬件,最理想的调试方式是将程序上传到物理设备上运行。但在没有物理设备的情况下,也可以使用 Android 模拟器提供的虚拟方式模拟设备的位置变化,调试具有位置服务的应用程序。

首先打开 DDMS 中的模拟器控制,在 Location Controls 中的 Longitude 和 Latitude 部分输入设备当前的经度和纬度,然后点击 Send 按钮,就将虚拟的位置信息发送到 Android 模拟器中。如图 17-2 所示。

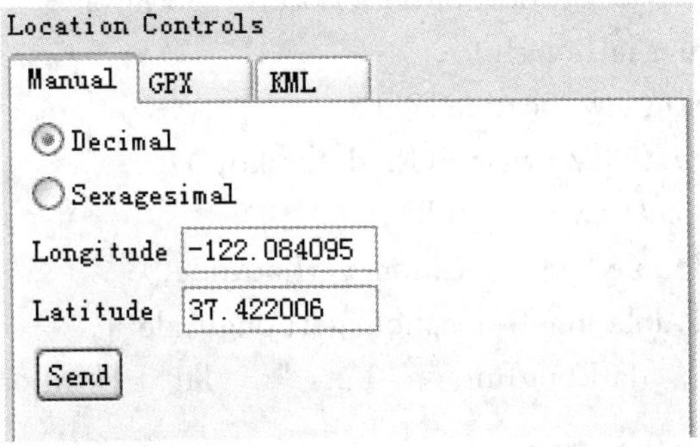

图 17-2 模拟器控制

在程序运行过程中,可以在模拟器控制器中改变经度和纬度坐标值,程序在检测到位置的变化后,会将最新的位置信息显示在界面上。但作者在 1.5 版本的 Android 模拟器中进行调试时,发现模拟器控制器只能成功地将虚拟坐标发送到模拟器中 2 次,超过 2 次后模拟器对新发送的虚拟坐标不再响应。

下面是 CurrentLocationDemo 示例中 LocationBasedServiceDemo.java 文件的完整代码。

```
1. package edu.xzceu.LocationBasedServiceDemo;
2. import android.app.Activity;
3. import android.content.Context;
4. import android.os.Bundle;
5. import android.widget.TextView;
6. import android.location.Location;
7. import android.location.LocationListener;
8. import android.location.LocationManager;
9. public class LocationBasedServiceDemo extends Activity {
10.     @Override
11.     public void onCreate(Bundle savedInstanceState) {
12.         super.onCreate(savedInstanceState);
13.         setContentView(R.layout.main);
14.         String serviceString = Context.LOCATION_SERVICE;
15.         LocationManager locationManager =
              (LocationManager)getSystemService(serviceString);
16.         String provider = LocationManager.GPS_PROVIDER;
```

```java
17.            Location location =
    locationManager.getLastKnownLocation(provider);
18.            getLocationInfo(location);
19.            locationManager.requestLocationUpdates(provider, 2000, 0,
    locationListener);
20.     }
21.     private void getLocationInfo(Location location){
22.         String latLongInfo;
23.         TextView locationText =
    (TextView)findViewById(R.id.txtshow);
24.         if (location != null){
25.           double lat = location.getLatitude();
26.           double lng = location.getLongitude();
27.             latLongInfo = "Lat: " + lat + "\nLong: " + lng;
28.         }
29.         else{
30.             latLongInfo = "No location found";
31.         }
32.         locationText.setText("Your Current Position is:\n" +
    latLongInfo);
33. }
34.
35.     private final LocationListener locationListener = new LocationListener(){
36.         @Override
37.         public void onLocationChanged(Location location) {
38.             getLocationInfo(location);
39.         }
40.         @Override
41.         public void onProviderDisabled(String provider) {
42.             getLocationInfo(null);
43.         }
44.         @Override
45.         public void onProviderEnabled(String provider) {
46.             getLocationInfo(null);
47.         }
48.         @Override
49.         public void onStatusChanged(String provider, int status, Bundle
    extras) {
50.         }
51.     };
52. }
```

17.2　Google 地图应用

17.2.1　申请地图密钥

首先向 Google 申请一组经过验证的"地图密钥"（Map API Key），然后使用 MapView（com.google.android.maps.MapView）就可以将 Google 地图嵌入到 Android 应用程序中，才能正常使用 Google 的地图服务。"地图密钥"是访问 Google 地图数据的密钥，无论是模拟器还是在真实设备中需要使用这个密钥。

注册"地图密钥"的第一步是申请一个 Google 账户，也就是 Gmail 电子邮箱，申请地址是 https://www.google.com/accounts/Login。找到保存 Debug 证书的 keystore 的保存位置，并获取证书的 MD5 散列值。keystore 是一个密码保护的文件，用来存储 Android 提供的用于调试的证书。获取 MD5 散列值的主要目的是为下一步申请"地图密钥"做准备，获取证书的保存地址：首先打开 Eclipse，通过 Window → Preferences 打开配置窗体，在 Android → Build 栏中的 Default debug keystore 中可以找到。如图 17-3 所示。

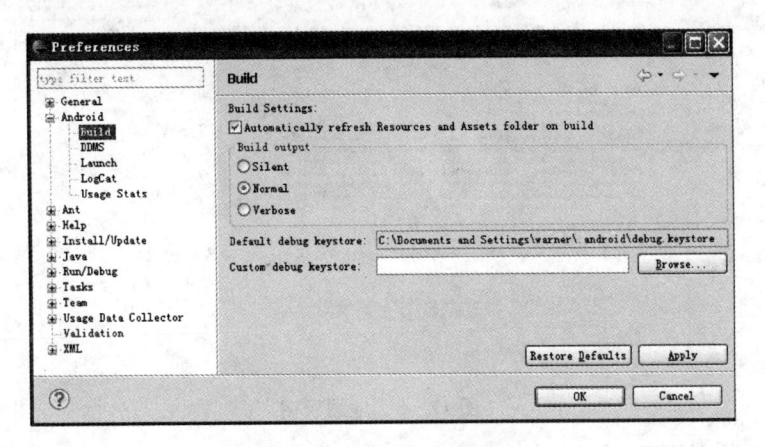

图 17-3　获取证书的保存地址

为了获取 Debug 证书 MD5 散列值的，需要打开命令行工具 CMD，然后切换到 keystore 的目录，输入如下命令：

```
keytool - list - keystore debug.keystore
```

如果提示无法找到 keytool，可以将＜Java SDK＞/bin 的路径添加到系统的 PATH 变量中，在提示输入 keystore 密码时，输入缺省密码 android，MD5 散列将显示在最下方。笔者的 MD5 散列值为：68:76:89:C8:A4:24:61:F9:EA:F3:F7:70:CC:FD:C8:15。如图 17-4 所示。

图 17-4 获取 Debug 证书 MD5 散列值

申请"地图密钥"的最后一步是打开申请页面,输入 MD5 散列值申请页面的地址是:http://code.google.com/intl/zh-CN/android/add-ons/google-apis/maps-api-signup.html。如图 17-5 所示。

图 17-5 地图密钥

输入 MD5 散列值后,点击 Generate API Key 按钮,将提示用户输入 Google 账户。正确输入 Google 账户后,将产生申请"地图密钥"的获取结果。

作者获取的"地图密钥"是 0mVK8GeO6WUz4S2F94z52CIGSSlvlTwnrE4 * * * *,在以后使用到 MapView 的时候都需要输入这个密钥。但需要注意的是,除作者本人外并不能使用这个密钥,需要根据自己的 Debug 证书的 MD5 散列值,重新到 Google 网站上申请一个用于调试程序的"地图密钥"。

17.2.2 使用 Google 地图

MapView 是地图的显示控件,可以设置不同的显示模式,例如卫星模式、街道模式或交通模式。MapController 则是 MapView 的控制器,可以控制 MapView 的显示中心和缩放级别等功能。

下面以 GoogleMapDemo 为例,说明如何在 Android 系统中开发 Google 地图程序。这个示例将在程序内部设置一个坐标点,然后在程序启动时,使用 MapView 控件在地图上显示这个坐标点的位置。

在建立工程时将 com. google. android. maps 的扩展库添加到工程中,这样就可以使用 Google 地图的所有功能。添加 com. google. android. maps 扩展库的方式是在创建工程时,在 Build Target 项中选择 Google APIs。如图 17-6 所示。

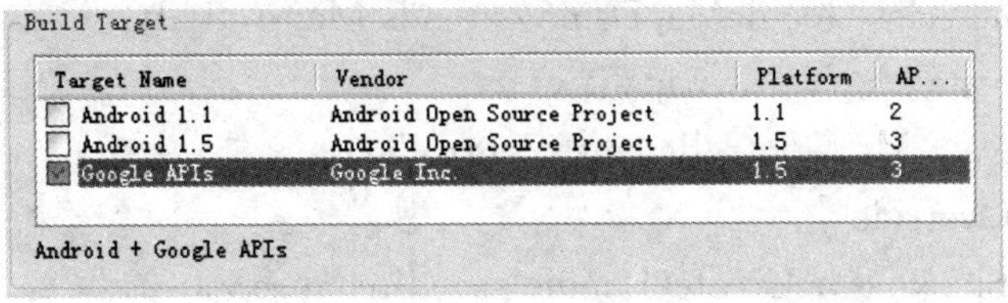

图 17-6 添加 com. google. android. maps 扩展库

创建工程后,修改/res/layout/main. xml 文件,在布局中加入一个 MapView 控件,并设置刚获取的"地图密钥"。

main. xml 文件的完整代码如下:

```
1.  <?xml version="1.0" encoding="utf-8"?>
2.  <LinearLayout xmlns:android=" http://schemas.android.com/apk/res/
    android"
3.    android:orientation="vertical"
4.    android:layout_width="fill_parent"
5.    android:layout_height="fill_parent">
6.    <TextView android:layout_width="fill_parent"
7.      android:layout_height="wrap_content"
8.      android:text="@string/hello"/>
9.    <com.google.android.maps.MapView
10.     android:id="@+id/mapview"
11.     android:layout_width="fill_parent"
12.     android:layout_height="fill_parent"
13.     android:enabled="true"
14.     android:clickable="true"
15.     android:apiKey="0mVK8GeO6WUz4S94z52CIGSSlvlTwnrE4DsiA"/>
16. </LinearLayout>
```

仅在布局中添加 MapView 控件,还不能够直接在程序中调用这个控件,还需要将程序本身设置成 MapActivity(com. google. android. maps. MapActivity)。MapActivity 类负责处理显示 Google 地图所需的生命周期和后台服务管理。

下面先给出整个 GoogleMapDemo. java 文件的完整代码:

```
1.  package edu.xzceu.GoogleMapDemo;
2.  import com.google.android.maps.GeoPoint;
3.  import com.google.android.maps.MapActivity;
4.  import com.google.android.maps.MapController;
5.  import com.google.android.maps.MapView;
6.  import android.os.Bundle;
7.  public class GoogleMapDemo extends MapActivity {
8.    private MapView mapView;
9.    private MapController mapController;
10.   @Override
11.   public void onCreate(Bundle savedInstanceState) {
12.       super.onCreate(savedInstanceState);
13.       setContentView(R.layout.main);
14.       mapView = (MapView)findViewById(R.id.mapview);
15.       mapController = mapView.getController();
16.       Double lng = 126.676530486 * 1E6;
17.       Double lat = 45.7698895661 * 1E6;
18.       GeoPoint point = new GeoPoint(lat.intValue(), lng.intValue());
19.       mapController.setCenter(point);
20.       mapController.setZoom(11);
21.       mapController.animateTo(point);
22.       mapView.setSatellite(false);
23.   }
24.   @Override
25.   protected boolean isRouteDisplayed() {
26.       // TODO Auto-generated method stub
27.       return false;
28.   }
29. }
```

第 20 行代码获取了 MapController。第 22 行和第 23 行代码设定的经度为 126.676530486 * 1E6、纬度为 45.7698895661 * 1E6 的地理坐标点。第 26 行代码将这个坐标转化为 GeoPoint 再使用。第 26 行代码设置 MapView 的"显示中点"。第 27 行代码设置放大层级。在第 28 行代码将 MapView 显示区域的中心移动到第 26 行设置的"显示中心"。第 30 行代码是设定 MapView 的地图显示模式是否为卫星模式，设置 true 则为卫星模式，设置 false 则为普通模式。第 34 行代码 isRouteDisplayed() 方法，是用来统计程序是否显示在 Google 地图中显示路径信息，默认为不显示。

由于获取 Google 地图是需要使用互联网的，所以在运行前还需要在 AndroidManifest.xml 文件中，添加允许访问互联网的许可。AndroidManifest.xml 文件的完整代码如下：

```
1. <? xml version="1.0" encoding="utf-8"?>
2. <manifest xmlns:android="http://schemas.android.com/apk/res/android"
3.    package="edu.xzceu.GoogleMapDemo"
4.    android:versionCode="1"
5.    android:versionName="1.0">
6.      <application android:icon="@drawable/icon" android:label="@string/app_name">
7.          <activity android:name=".GoogleMapDemo"
8.              android:label="@string/app_name">
9.              <intent-filter>
10.                 <action android:name="android.intent.action.MAIN" />
11.                 <category android:name="android.intent.category.LAUNCHER" />
12.             </intent-filter>
13.         </activity>
14.         <uses-library android:name="com.google.android.maps"></uses-library>
15.     </application>
16.     <uses-sdk android:minSdkVersion="3" />
17.     <uses-permission android:name="android.permission.INTERNET"/>
18. </manifest>
```

地图模式的运行结果如图 17-7 所示。卫星模式的运行结果如图 17-8 所示。

图 17-7　地图模式

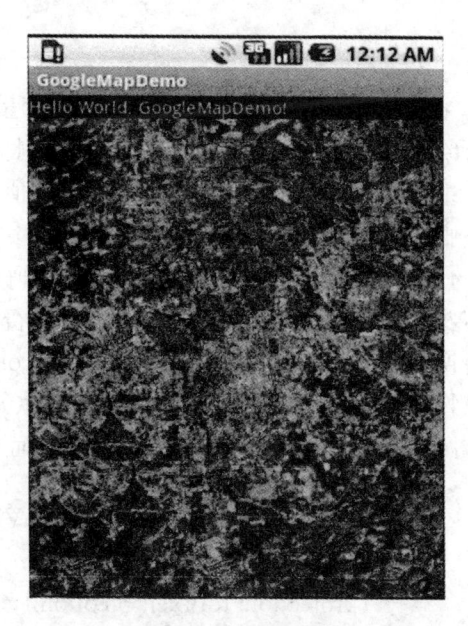

图 17-8　卫星模式

17.2.3 使用 Overlay

通过在 MapView 上添加覆盖层,可以在指定的位置加添加注解、绘制图像或处理进行鼠标事件等。Google 地图上可以加入多个覆盖层,所有覆盖层均都在地图图层之上,每个覆盖层均可以对用户的点击事件做出响应。

创建覆盖层继承 Overlay 类的子类,并通过重载 draw()方法为指定位置添加注解,重载 onTap()方法处理用户的点击操作。下面的代码是创建 Overlay 的最小代码集合:

```
1. public class TextOverlay extends Overlay {
2.    @Override
3.    public void draw(Canvas canvas, MapView mapView, boolean shadow) {
4.        if (shadow == false){
5.        }
6.        else{
7.        }
8.        super.draw(canvas, mapView, shadow);
9.    }
10.
11.   @Override
12.   public boolean onTap(GeoPoint p, MapView mapView) {
13.       return false;
14.   }
15. }
```

第 3 行代码中 draw()方法,shadow 变量是用来区分绘制图层的。false 表示在覆盖层上进行绘制,true 则表示在隐藏层上进行绘制。

第 14 行代码是 onTap()方法的返回值。返回 false 表示覆盖层不处理点击事件,返回 true 则表示已经处理了点击事件。

在覆盖层绘制图形或文字需要使用"画布"(Canvas)来实现,绘制的位置是屏幕坐标,这就需要将地图上的物理坐标与屏幕坐标进行转换。Projection 类提供了物理坐标和屏幕坐标的转换功能,可在经度和纬度表示的 GeoPoint 点和屏幕上 Point 点进行转换。toPixels()方法将物理坐标转换为屏幕坐标,fromPixels()方法将屏幕坐标转换为物理坐标。

两个方法的具体使用方法可以参考下面的代码:

```
1. Projection projection = mapView.getProjection();
2.
3. projection.toPixels(geoPoint, point);
4. projection.fromPixels(point.x, point.y);
```

下面的内容以 MapOverlayDemo 示例,说明如何在 Google 地图上添加覆盖层,并在预订的物理坐标上显示提示信息。图 17-9 是 MapOverlayDemo 示例的运行结果。

图 17-9　MapOverlayDemo 示例的运行结果

TextOverlay 类是 MapOverlayDemo 示例的覆盖层，主要重载了 draw() 方法，在指定的物理坐标上绘制了标记点和提示文字。TextOverlay.java 文件的核心代码如下。

```
1. public class TextOverlay extends Overlay {
2.    private final int mRadius = 5;
3.
4.    @Override
5.    public void draw(Canvas canvas, MapView mapView, boolean shadow) {
6.       Projection projection = mapView.getProjection();
7.
8.       if (shadow == false){
9.          Double lng = 31.938069 * 1E6;
10.         Double lat = 118.89628 * 1E6;
11.         GeoPoint geoPoint = new GeoPoint(lat.intValue(), lng.intValue());
12.
13.         Point point = new Point();
14.         projection.toPixels(geoPoint, point);
15.
16.         RectF oval = new RectF(point.x - mRadius, point.y - mRadius, point.x + mRadius, point.y + mRadius);
17.
18.         Paint paint = new Paint();
19.         paint.setARGB(250, 250, 0, 0);
```

```
20.            paint.setAntiAlias(true);
21.            paint.setFakeBoldText(true);
22.
23.        canvas.drawOval(oval,paint);
24.            canvas.drawText("标记点",point.x+2*mRadius,point.y,paint);
25.        }
26.        super.draw(canvas,mapView,shadow);
27. }
28.
29.    @Override
30.    public boolean onTap(GeoPoint p,MapView mapView){
31.        return false;
32.    }
33. }
```

第 2 行代码定义了绘制半径变量 mRadius,供定义绘制范围使用。第 14 行代码使用 Projection 完成了从物理坐标到屏幕坐标的转换。第 16 行代码 oval 设定标记点的大小。第 19 行设置了绘制颜色。第 20 行开启了平滑设置,防止文字出现锯齿。第 23 行代码绘制了圆形的标记点。第 24 行代码绘制了提示文字,第 2 个和第 3 个参数是绘制屏幕的 x 坐标和 y 坐标。

建立了覆盖层后,还需要把覆盖层添加到 MapView 上。MapOverlayDemo.java 的核心代码如下:

```
1. public class MapOverlayDemo extends MapActivity {
2.    private MapView mapView;
3.    private MapController mapController;
4.    private TextOverlay textOverlay;
5.    @Override
6.    public void onCreate(Bundle savedInstanceState) {
7.        super.onCreate(savedInstanceState);
8.        setContentView(R.layout.main);
9.            mapView=(MapView)findViewById(R.id.mapview);
10.                mapController=mapView.getController();
11.
12.            Double lng=31.938069*1E6;
13.            Double lat=118.89628*1E6;
14.        GeoPoint point=new GeoPoint(lat.intValue(),lng.intValue());
15.
16.        mapController.setCenter(point);
```

```
17.        mapController.setZoom(11);
18.        mapController.animateTo(point);
19.
20.        textOverlay = new TextOverlay();
21.        List<Overlay> overlays = mapView.getOverlays();
22.        overlays.add(textOverlay);
23.    }
24.
25.    @Override
26.    protected boolean isRouteDisplayed() {
27.        return false;
28.    }
29. }
```

第 22 行代码实例化了 TextOverlay 对象。第 23 行代码通过 getOverlays() 方法，获取 MapView 已有的覆盖层。第 24 行代码使用 add() 方法将 TextOverlay 对象添加到 MapView 中。

习题与思考题

1. 讨论位置服务和地图应用的发展前景。

2. 编程实现轨迹追踪软件。每间隔 60 秒，同时距离移动大于 1 米的情况下，记录一次位置信息，在 Google 地图上绘制 600 秒的行动轨迹。

第 18 章

Android NDK 开发

☆18.1　NDK 简介
☆18.2　NDK 编译环境
☆18.3　NDK 开发示例

本章学习目标：了解 Android NDK 的用途；掌握 Android NDK 编译环境的安装与配置方法；掌握 Android NDK 的开发方法。

18.1　NDK 简介

Android NDK(Android Native Development Kit)是一系列的开发工具,允许程序开发人员在 Android 应用程序中嵌入 C/C++语言编写的非托管代码。

Android NDK 优势：

(1) 解决了核心模块使用托管语言开发执行效率低下的问题。

(2) 允许程序开发人员直接使用 C/C++源代码,极大地提高了 Android 应用程序开发的灵活性。

Android NDK 不足：

(1) 增加程序的例如信号处理或物理仿真,使用非托管代码运行效率最高,复杂性。

(2) 增加了程序调试的难度。

(3) CPU 运算量大和内存消耗较少的部分(因此并不是所有的核心部分都适合使用)。

(4) C/C++语言编写。

Android NDK 的版本是 r6b,集成了交叉编译器,支持 ARMv5TE 处理器指令集、JNI 接口和一些稳定的库文。

Android NDK 提供一系列的说明文档、示例代码和开发工具,指导程序开发人员使用 C/C++语言进行库文件开发,并提供便捷工具将库文件打包到 apk 文件中。

18.2　NDK 编译环境

Android NDK 编译环境支持 Windows、Linux 和 MacOS,本书仅介绍 Windows 系统的编译环境配置方法。

1. 下载 Android NDK 的安装包

将下载的 ZIP 文件解压缩到用户的 Android 开发目录中,作者将 Android NDK 解压到 E:\Android 目录中,ZIP 文件中包含一层目录,因此 Android NDK 的最终路径为 E:\Android\android‐ndk‐r6b。

2. 下载并安装 Cygwin

Android NDK 目前还不支持在 Windows 系统下直接进行交叉编译,因此需要在 Windows 系统下安装一个 Linux 的模拟器环境 Cygwin,完成 C/C++代码的交叉编译工作。

Android NDK 要求 GNU Make 的版本高于或等于 3.18,之前的版本并没有经过测试,因此需要安装较新版本的 Cygwin。

Cygwin 的最新版本可以到官方网站 http://www.cygwin.com 下载,也可以到中文的映像网站 http://www.cygwin.cn 下载。

如图 18‐1,在 Cygwin 的安装过程中,需要将 Devel 下的 gcc 和 make 的相关选项选上,否则 Cygwin 将无法编译 C/C++代码文件。

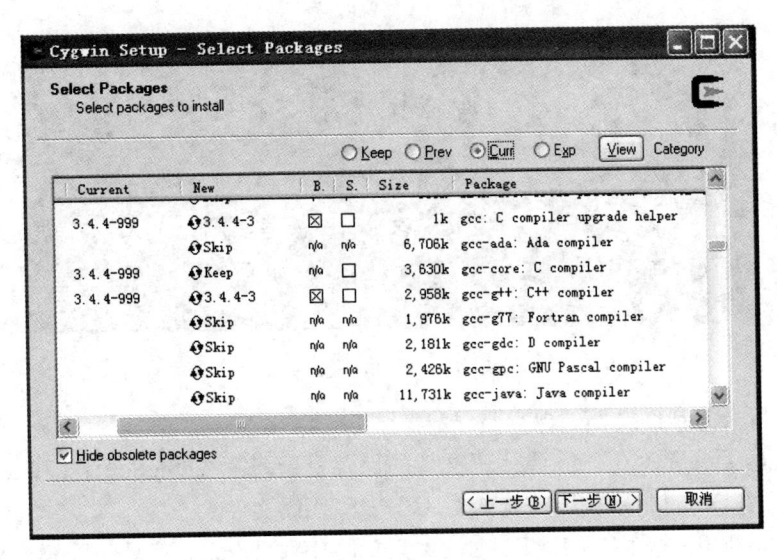

图 18-1　Cygwin 安装过程

3. 配置 Cygwin 的 NDK 开发环境

在缺省情况下，Cygwin 安装在 C 盘的根目录下，修改 C:\cygwin\home\username\.bash_profile 文件，username 会根据用户使用的用户名称而变化。

在.bash_profile 文件的结尾处添加如下代码：

ANDROID_NDK_ROOT=/cygdrive/e/android/android-ndk-1.5_r1

export ANDROID_NDK_ROOT

上面的代码说明了 Android NDK 所在的目录是 e 盘 android/android-ndk-1.5_r1。

如果 Android NDK 安装在 c 盘的 TestAndroid/android-ndk-1.5_r1 中，则上面的代码应为：

```
ANDROID_NDK_ROOT=/cygdrive/e/android/android-ndk-1.5_r1
export ANDROID_NDK_ROOT
```

```
ANDROID_NDK_ROOT=/cygdrive/c/TestAndroid/android-ndk-1.5_r1
export ANDROID_NDK_ROOT
```

4. 测试开发环境

（1）首先启动 Cygwin，然后切换到<Android NDK>/build 目录中，运行 host-setup.sh 文件。

（2）如果运行结果如图 18-2 所示，说明 Android NDK 的开发环境已经可以正常工作了。

（3）Android NDK 的编译环境已经安装配置完毕。

5. Android NDK 的目录结构

在 android-ndk-2.1_r1 目录中，包含 5 个子目录和 2 个文件。

（1）samples 目录是 Android 工程的保存目录。

（2）build 目录保存了交叉编译工具、编译脚本和配置文件。

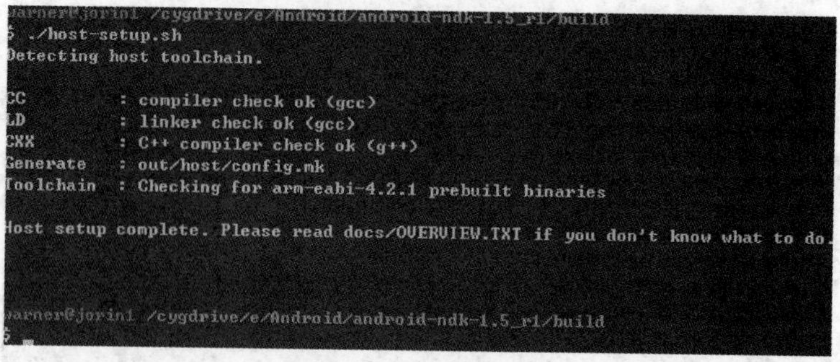

图 18-2　运行环境

（3）docs 目录是帮助文档的保存目录。

（4）sources 目录是 C/C++源代码文件的保存目录，其下的 hello-jni 和 tow-libs 子目录，分别保存了 NDK 自带示例所需要的 C/C++源代码文件。

（5）GNUmakfile 文件和 README.TXT 文件分别是 make 工具的配置文件和 NDK 的说明文件。

docs 目录中的帮助文件说明，如表 18-1 所示。

表 18-1　docs 目录文档

文件名	说　明
INSTALL.TXT	NDK 的安装与配置说明文档
OVERVIEW.TXT	NDK 的用途和使用范围的说明文档
ANDROID—MK.TXT	Android.mk 文件的说明文档，Android.mk 文件定义了需要编译的 C/C++源代码
APPLICATION—MK.TXT	Application.mk 文件的说明文档，Application.mk 文件定义了 Android 工程需要调用的 C/C++源代码
HOWTO.TXT	关于 NDK 开发的一般性信息
SYSTEM—ISSUES.TXT	使用 NDK 开发时所需要注意的问题
STABLE—APIS.TXT	NDK 头文件所支持的稳定的 API 类表

Android NDK 自带两个示例 hello-jni 和 tow-libs。

hello-jni 是一个非常简单的例子，非托管代码实现了一个可以返回字符串的共享库，Android 工程调用这个共享库获取字符串，然后显示在用户界面上。

tow-libs 是稍微复杂一些的例子，使用非托管代码实现了一个数学运算的共享库，Android 工程动态加载这个共享库，并调用其中的函数，函数功能是通过使用静态库实现的。

18.3　NDK 开发示例

在进行 NDK 开发时，一般需要同时建立 Android 工程和 C/C++工程，然后使用 NDK 编译 C/C++工程，形成可以被调用的共享库，最后共享库文件会被拷贝到 Android 工程中，并被直接打包到 apk 文件中。

后面的内容将 AndroidNdkDemo 示例说明如何进行 Android NDK 开发。

AndroidNdkDemo 是一个进行加法运算的示例，程序会随机产生两个整数，然后调用 C 语言开发的共享库对这两个整数进行加法运算，最后将运算结果显示在用户界面上。如图 18-3 所示。

图 18-3　AndroidNdkDemo 示例的界面

进行 Android NDK 开发一般要经过如下的步骤：
(1) 建立 Application.mk 文件。
(2) 建立 Android 工程。
(3) 建立 Android.mk 文件。
(4) 建立 C 源代码文件。
(5) 编译共享库模块。
(6) 建立 Application.mk 文件。

1. 建立 Application.mk 文件

(1) 在 apps 目录中建立应用程序目录，AndroidNdkDemo 示例的应用程序目录为 ndk-demo。

(2) 在 ndk-demo 目录中建立一个空目录 project，这个目录以后会用来存放 Android 工程。

(3) 在 ndk-demo 目录中建立一个名为 Application.mk 的文件，用来描述 Android 工程将调用的共享库。

AndroidNdkDemo 示例的目录结构如图 18-4 所示。

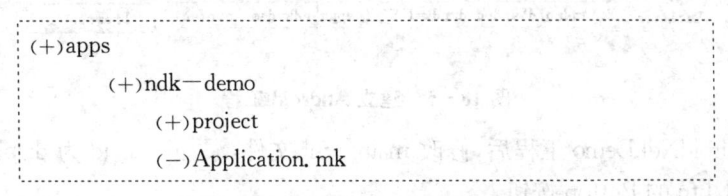

```
(+)apps
    (+)ndk-demo
        (+)project
        (-)Application.mk
```

图 18-4　AndroidNdkDemo 目录结构

在进行 NDK 开发时，在应用程序目录中一定要有 Application.mk 文件，用来声明 Android 工程需要调用的非托管模块（如静态库或共享库）。

AndroidNdkDemo 示例的 Application.mk 的代码如下：

```
1. APP_PROJECT_PATH := $(call my-dir)/project
2. APP_MODULES := add-module
```

第1行的变量 APP_PROJECT_PATH 表示 Android 工程所在的目录,在生产共享库文件后,APK 将自动将共享库文件拷贝到<app>\libs\armeabi 目录中。本示例将共享库文件拷贝到 apps\ndk-demo\project\libs\armeabi 目录中。

第2行代码中的变量 APP_MODULES 表示 Android 工程需要调用的非托管模块,如果存在多个非托管模块,使用空格进行分隔。本示例调用的非托管模块为 add-module,对应在后面涉及的 Android.mk 文件。Application.mk 的变量说明如表 18-2 所示。

表 18-2 变量说明

变量	强制使用	说明
APP_PROJECT_PATH	是	Android 工程所在的目录
APP_MODULES	是	Android 工程需要调用的非托管模块
APP_OPTIM	否	指定优化等级,包含两个等级 debug 和 release,release 是缺省设置。debug 生产非优化代码,更加易于调试
APP_CFLAGS	否	编译 C 代码时所传递的编译器标志
APP_CXXFLAGS	否	编译 C++ 代码时所传递的编译器标志
APP_CPPFLAGS	否	编译 C/C++ 代码时所传递的编译器标志

2. 建立 Android 工程

如图 18-5 所示,在 project 目录中建立 Android 工程时,需要取消复选框"Use default location",并指定预先建立的 project 文件夹作为工程文件夹。

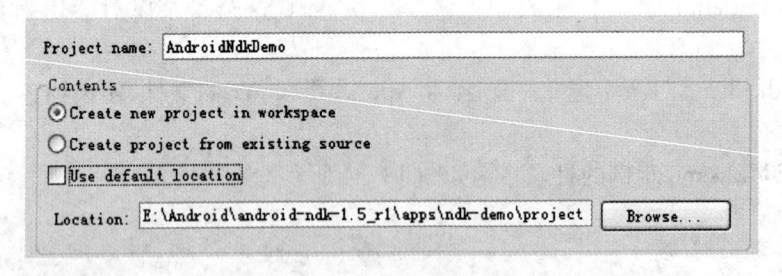

图 18-5 建立 Android 工程

在建立 AndroidNdkDemo 工程后,修改 main.xml 文件,添加一个 id 为 display 的 TextView 和一个 id 为 add_btn 的 Button 按钮。

程序中的生产随机数和调用的代码在 AndroidNdkDemo.java 文件中,AndroidNdkDemo.java 文件的核心代码如下所示:

```
1. public class AndroidNdkDemo extends Activity {
2.     @Override
```

```
3.   public void onCreate(Bundle savedInstanceState) {
4.      super.onCreate(savedInstanceState);
5.      setContentView(R.layout.main);
6.      final TextView displayLable = (TextView)findViewById(R.id.display);
7.      Button btn = (Button)findViewById(R.id.add_btn);
8.      btn.setOnClickListener(new View.OnClickListener(){
9.                      @Override
10.             public void onClick(View v) {
11.                 double randomDouble = Math.random();
12.                 long x = Math.round(randomDouble * 100);
13.                 randomDouble = Math.random();
14.                 long y = Math.round(randomDouble * 100);
15.
16.                 //System.loadLibrary("add-module");
17.                 long z = add(x, y);
18.                 String msg = x+" + "+y+" = "+z;
19.                 displayLable.setText(msg);
20.             }
21.         });
22.     }
23.     //public native long add(long x, long y);
24.
25.     public long add(long x, long y){
26.         return x+y;
27.     }
28. }
```

上面的代码有一个 NDK 开发的小技巧，在开发 C/C++ 的共享库前，可以使用具有相同和相近功能的 Java 函数进行替代。在代码第 17 行本应该调用共享库的 add() 函数，但为了便于开发和调试，在代码第 25 行到第 27 行，使用 Java 代码开发了一个功能相同的 add() 函数。这样即使在没有完成 C/C++ 的共享库开发前，也可以对这个 Android 工程进行界面部分的调试。

第 16 行和第 23 行注释掉的代码，就是在 C/C++ 的共享库开发完毕后需要使用的代码。其中第 16 行是动态加载共享库的代码，加载的共享库名称为 add-module。第 23 行用来声明共享库的函数，使用 C/C++ 开发的共享库必须有同名的函数。在共享库开发完毕后，取消第 16 行和第 23 行代码的注释，并注释掉第 25 行到第 27 行代码，这样程序就可以正常调用共享库内的函数进行加法运算。

3. 建立 Android.mk 文件

建立 C/C++ 源代码文件前，首先需要在 sources 目录中建立模块目录，AndroidNdkDemo 示例的模块目录为 add-module，这个模块目录的名称与 Application.mk 文件中声明的模块名称相同。

add-module 目录中包含两个文件，Android.mk 和 add-module.c，目录结构如图 18-6 所示。

```
(+)sources
    (+)add-module
        (-)Android.mk
        (-)add-module.c
```

图 18-6 add-module 目录结构

Android.mk 是为 NKD 编译系统准备的脚本文件，用来描述模块需要编译 C/C++文件的信息。

通常 NKD 编译系统会搜寻＄NDK/sources/＊/目录中的所有 Android.mk 文件，但如果程序开发人员将 Android.mk 文件放置在下一级目录中，则需要在上一级目录中的 Android.mk 文件添加如下代码：

```
include $(call all-subdir-makefiles)
```

下面来分析 AndroidNdkDemo 示例的 add-module 模块的 Android.mk 文件，Android.mk 文件的代码如下：

```
1. LOCAL_PATH := $(call my-dir)
2.
3. include $(CLEAR_VARS)
4.
5. LOCAL_MODULE    := add-module
6. LOCAL_SRC_FILES := add-module.c
7.
8. include $(BUILD_SHARED_LIBRARY)
```

每个 Android.mk 文件都必须以第 1 行代码开始，变量 LOCAL_PATH 用来定义需要编译的 C/C++源代码的位置，my-dir 由 NKD 编译系统提供，表示当前目录的位置。代码第 3 行的 include ＄(CLEAR_VARS)表示清空所有以 LOCAL_开始的变量，例如 LOCAL_MODULE、LOCAL_SRC_FILES、LOCAL_STATIC_LIBRARIES 等，但第 1 行定义的 LOCAL_PATH 不在清空的范围内。因为所有的脚本都将粘贴到同一个 GNU Make 的执行上下文中，而且所有变量都是全局变量，因此必须在每次使用前清空所有以前用过的变量。

第 5 行代码变量 LOCAL_MODULE 用来声明模块名称，模块名称必须唯一，而且中间不能存在空格。NDK 编译系统将会在模块名称前自动添加 lib 前缀，然后生产 so 文件。这里的模块名称为 add-module，生产的共享库文件名为 libadd-module.so。但需要注意的是，如果程序开发人员使用具有 lib 前缀的模块名称，NDK 编译系统将不再添加前缀，例如模块名称为 libsub，生产的共享库文件名为 libsub.so。

第 6 行代码中的变量 LOCAL_SRC_FILES 表示编译模块所需要使用的 C/C++ 文件列表,但不需要给出头文件的列表,因为 NKD 编译系统会自动计算依赖关系。add-module 模块仅需要一个 C 文件,文件名为 add-module.c。缺省情况下,结尾名为.c 的文件是 C 语言源文件,结尾名为.cpp 的文件是 C++ 语言源文件。

第 8 行代码 include $(BUILD_SHARED_LIBRARY)表示 NKD 编译系统构建共享库,如果变量 BUILD_SHARED_LIBRARY 更改为 BUILD_STATIC_LIBRARY,则表示需要 NKD 编译系统构建静态库。

4. 建立 C 源代码文件

根据 Android.mk 文件的声明,add-module 模块仅包含一个 C 源代码文件 add-module.c。add-module.c 文件的作用是实现两个整数加法运算功能,全部代码如下:

```
1. #include <jni.h>
2. jlong Java_edu_xzceu_AndroidNdkDemo_AndroidNdkDemo_add( JNIEnv *
   env, jobject this, jlong    x, jlong    y)
3. {
4.     return x+y;
5. }
```

第 1 行代码引入的是 JNI(Java Native Interface)的头文件。第 3 行代码是函数名称,jlong 表示 Java 长型整数,Java_edu_xzceu_AndroidNdkDemo_AndroidNdkDemo_add 的构成为 Java_<包名称>_<类>_<函数>,其中<函数>的名称和参数要与 AndroidNdkDemo.java 文件定义的函数一致,AndroidNdkDemo.java 文件定义的函数为 public native long add (long x, long y)。第 5 行代码用来返回加法运算结果。

5. 编译共享库模块

首先启动 cygwin,然后切换到 Androd NDK 的主目录下,键入如下的编译命令。

```
make APP=ndk-demo
```

ndk-demo 是 apps 目录下的应用程序目录名称。在指定应用程序(目录)名称后,NKD 编译系统会首先找到目录中的 Application.mk 文件,根据 Application.mk 文件的信息,确定该 Android 共享需要使用 add-module 模块;然后在 sources 目录中搜索所有 Android.mk 文件,在找到与 add-module 模块匹配的 Android.mk 文件后,根据 Android.mk 文件提供的信息编译指定的 C/C++ 源代码文件,形成共享库文件;最后将生产的共享库文件拷贝到 Android 工程的指定目录中。

如图 18-7 所示,提示信息包括编译 add-module 模块所使用到的文件,生产 so 文件的文件名和 so 文件的安装位置。为了确认是否成功编译了模块,用户可以打开 apps/ndk-demo/project/libs/armeabi 目录,如果目录中存在 libadd-module.so 文件,则表示编译成功。

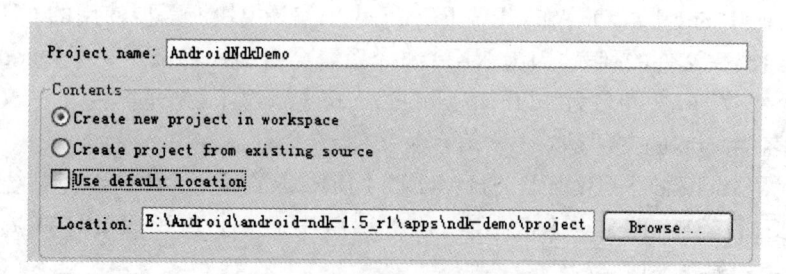

图 18－7　编译成功的提示信息

6. 运行 Android 程序

在运行 AndroidNdkDemo 示例程序前，务必将 AndroidNdkDemo.java 文件中第 16 行和第 23 行的注释取消，并注释掉第 25 行到第 27 行代码。

习题与思考题

1. 简述 Android NDK 开发的优势和不足。
2. 说明 Android NDK 应用程序开发的一般步骤。
3. 参考 NDK 的 tow－libs 示例，使用静态库实现 AndroidNdkDemo 示例中加法运算的函数功能。
4. 使用 NDK 能够提高复杂函数的运算速度，但程序运行效率的提升并不容易度量。分别使用 C/C++和 Java 语言设计一个具有复杂运算的函数，通过对比函数的调用和返回时间，分析 NDK 对提高程序运行效率的能力。

参考文献

[1] (U.S.A.)Gabriel Svennerberg. Beginning Google Maps API 3[M]. APRESS,2010.

[2] (U.S.A.)John Eddy,Patricia DiGiacomo Eddy. Google on the Go：Using an Android-Powered Mobile Phone[M]. Que,2009.

[3] E2ECloud 工作室. 深入浅出 Google Android[M]. 北京：人民邮电出版社,2009.

[4] 靳岩,姚尚朗. Android 开发入门与实战[M]. 北京：人民邮电出版社,2009.

[5] 张利国,龚海平. Andriod 移动开发案例详解[M]. 北京：人民邮电出版社,2010.

[6] 宋光照,傅江如,刘世军等. 手机软件测试最佳实践[M]. 北京：电子工业出版社,2009.

[7] 和凌志,郭世平. 手机软件平台架构解析[M]. 北京：电子工业出版社,2009.

[8] 余志龙等. Google Android SDK 开发范例大全[D]. 北京：人民邮电出版社,2010.

[9] 李炜. Google Android 开发入门指南[M]. 北京：人民邮电出版社,2009.

图书在版编目（CIP）数据

基于Android的手机应用软件开发教程 / 包依勤主编. —南京：南京大学出版社，2012.12（2013.8重印）
应用型本科院校计算机类专业校企合作实训系列教材
ISBN 978-7-305-10923-2

Ⅰ. ①基… Ⅱ. ①包… ②陈… Ⅲ. ①移动电话机－游戏程序－程序设计-高等学校-教材 Ⅳ. ①TN929.53②TP311.5

中国版本图书馆CIP数据核字（2012）第301315号

出版发行	南京大学出版社
社　　址	南京市汉口路22号　　邮编　210093
网　　址	http://www.NjupCo.com
出版人	左　健
丛书名	应用型本科院校计算机类专业校企合作实训系列教材
书　　名	基于Android的手机应用软件开发教程
主　　编	包依勤
责任编辑	王秉华　单　宁　　编辑热线　025-83592146
照　　排	江苏南大印刷厂
印　　刷	南京人文印务有限公司
开　　本	787×1 092　1/16　印张 21.75　字数 560千
版　　次	2012年12月第1版　2013年8月第2次印刷
ISBN	978-7-305-10923-2
定　　价	45.00元
发行热线	025-83594756
电子邮箱	Press@NjupCo.com
	Sales@NjupCo.com(市场部)

* 版权所有，侵权必究
* 凡购买南大版图书，如有印装质量问题，请与所购图书销售部门联系调换